~ Energy
Saving House

of ELECOSAE, in association with the AA fund.

Thierry Salomon is an engineer and the founder of GEFOSAT, an association working on the development of renewable energy technologies and energy conservation. He has also been instrumental in the setting up of Ri2e an organisation promoting information resources on energy and the environment. Since 1991 they have opened over a dozen energy advice centres.

Stéphane Bedel is a heating engineer at GEFOSAT. He has managed energy advice services for over 10 years specialising in energy saving and renewables.

The Energy Saving House was originally published in France (as *La maison des [nega]watts*) by the publishing division of Terre Vivante, an eco-centre based in Mens. This edition has been translated and adapted for the UK by the Centre for Alternative Technology, Machynlleth, mid-Wales with the assistance of Heike Bauer (courtesy of The University of Aberystwyth Careers Service) and in association with Friends of the Earth.

The Energy Saving House

Thierry Salomon
&
Stéphane Bedel

Centre for
Alternative
Technology
Publications

**Friends of
the Earth**

First published in the UK in 2003.

Second edition updated and revised in 2007 by

The Centre for Alternative Technology Charity Ltd.

Registered Charity number: 265239

Machynlleth, Powys, SY20 9AZ, UK

• Tel. 01654 705980 • Fax. 01654 702782

• info@cat.org.uk

• www.cat.org.uk/catpubs

Cover illustration: Graham Preston

Illustrations: Terre Vivante (except where stated)

ISBN 978-1-902175-55-3

1 2 3 4 5 6 7 8 9 10

Mail Order copies from: Buy Green By Mail   Tel. 01654 705959

The details are provided in good faith and believed to be correct
at the time of writing, however no responsibility is taken for any
errors. Our publications are updated regularly; please let us know
of any amendments or additions which you think may be useful for
future editions.

Printed in Great Britain by The Cromwell Press, (01225 711400).

Produced on FSC approved 'Evolution Satin' paper and board

Acknowledgements

Thank you to everyone who helped us to create this practical guide to energy saving, following on from the exhibition of the same name at Terre Vivante: Sandrine Buresi, Danielle Suid, Renaud Mikolasek, David Lenoir, and Stéphane Davy (Gefosat and Ri2e), as well as Bernard Favre and Denis Delebecque (Louma Productions).

Thank you also to Laurent le Gruyader, Thomas Gueret, Bruno Peupartier, Alain Anglade, Yves Moch, Didier Cherel, Céline Vachey, Maxime Tassin...and to all those who provided us with vital information.

Finally, special thanks to to three inspirational pioneers: Claude Aubert, director of Terre Vivante, for his friendly and crucial re-reading; Olivier Sidler, unceasing in his search for negawatts, whose rigorous work has taught us to use energy better in our everyday lives; and Benoît Lebot, fervent promotor of energy efficiency in Europe and internationally.

As they have, we feel it is important to communicate and use relevant studies and statutory regulations to build a more energy responsible society. However, as with waste minimisation we all have a responsibility to put an end to energy inefficiency: this book invites you to play your part.

Thierry Salomon and Stéphane Bedel

'Tomorrow will not be as yesterday...
The future will not be as the past...

It will be new and depends on us: not so much a case of discovering the future but rather inventing it.' Gaston Berger.

The Energy Saving House was first published by the French eco-centre Terre Vivante, which mounted an exhibition 'La maison des negawatts' in 1999. The Centre for Alternative Technology (CAT) bought the English language rights to the book at the Frankfurt Bookfair in 2001.

Adapting the book to the English language and to British geography, climate and legislation has meant incorporating a number of changes and updates. We aimed to retain the look and tone of the original book, whilst ensuring that it was relevant to the widest possible audience. CAT now distributes books not only in Europe but also in North America and the rest of the world, via its website www.cat.org.uk/catpubs and hence we have tried to incorporate information on energy saving relevant to a broad range of circumstances. This inevitably means that there is some information included in the book which is exclusive to the UK, but also that there is information which is inappropriate for the British climate. The fields of renewable energy technology, energy conservation and environmental building are constantly developing and expanding and there are many opinions on best practice...not all the methods described in *The Energy Saving House* are wholly advocated by CAT staff. However, we hope that all those who read the book will find something within it that is useful to them and helps create lots of negawatts!

CAT Publications, 2007.

Contents

I. Introduction

What on earth is a 'negawatt'?

Energy: a squandered resource

Energy consumption reflects the inequalities between men

Without energy, there would be no life and no technical development. As we enter the 21st century, our planet is characterised by a deep divide between over-consumption and acute poverty: one single American citizen uses the equivalent of nearly 8 tonnes of oil each year, whereas an inhabitant of Bangladesh subsists on almost 50 times less! Electricity consumption is even more unequal: 1930.6kWh/person/year or 4677.2kWh/household/year of domestic electricity in the UK, compared to 10.2kWh/person/year in Ethiopia, with 37 per cent of the world's population having no access to electricity at all.[A]

Energy consumption across the world in tonnes of oil-equivalent per capita[B]

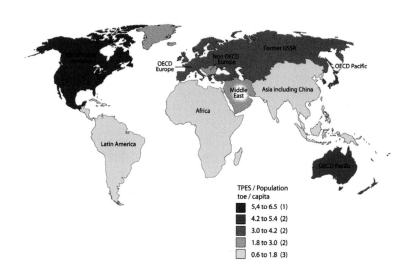

TPES / Population
toe / capita

- 5,4 to 6.5 (1)
- 4.2 to 5.4 (2)
- 3.0 to 4.2 (2)
- 1.8 to 3.0 (2)
- 0.6 to 1.8 (3)

The energy explosion

Reducing our energy requirement limits pollution

The world's energy consumption was stable for as long as humanity only used the energy it needed for survival and for its primary needs. From the 1850s on the Industrial Revolution has provoked an enormous increase in the demand for energy. Demand has continued to grow thanks to the combined effects of an increase in living standards and population growth. Currently, the world's energy demand grows by an average of two per cent each year. While this rate of growth has slowed down in the industrial nations, it is still increasing in developing countries where two-thirds of the world's population live.

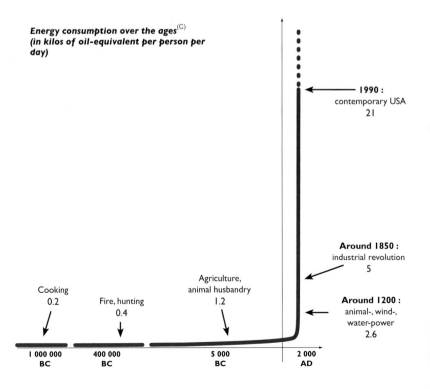

Energy consumption over the ages[C]
(in kilos of oil-equivalent per person per day)

1990 :
contemporary USA
21

Around 1850 :
industrial revolution
5

Agriculture,
animal husbandry
1.2

Around 1200 :
animal-, wind-,
water-power
2.6

Cooking
0.2

Fire, hunting
0.4

| 1 000 000 BC | 400 000 BC | 5 000 BC | 2 000 AD |

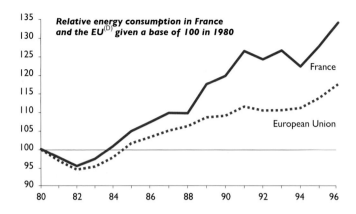

Relative energy consumption in France and the EU[D] given a base of 100 in 1980

France

European Union

Energy and pollution

The consumption of fossil fuels is one of the main reasons behind the destruction of the environment. The gases responsible for the so-called greenhouse effect are mostly released through the combustion of fossil fuels, industrial activities and deforestation.

Some of the gases that are used in refrigeration, and air-conditioning for buildings and cars, are responsible for the thinning of the ozone layer, which in turn lets dangerous UVB rays enter our atmosphere. These rays can have a harmful effect on both human health and the earth's ecosystem. However, CFCs are now banned in developed countries and HCFCs are being phased out in place of simple hydrocarbons.

Acid rain is a form of atmospheric pollution caused by oxides of sulphur and nitrogen. These gases are produced mainly by cars and factories, accumulate in rain clouds and fall as acid rain seriously damaging the earth's eco-systems.

The production of atomic energy presents an unprecedented risk for generations to come, since nuclear waste remains active for thousands of years. At present there is no satisfactory solution to the problem of what to do with nuclear waste. Simply burying it does not solve the problem.

Our generation has heedlessly gone on burning up non-renewable fuels

Our greed for energy and agricultural land is one of the main causes of deforestation leading to soil erosion. Population growth is back to extremely worrying levels, encouraging the loss of agricultural land to desertification as a direct result of initial deforestation and then over-cropping. This makes the likelihood of famine even greater.

Remember: resources are limited

If we continue to consume the same amount of energy, what resources will we have left tomorrow?

Oil will be the first resource to be exhausted, probably in about 2040, less than two generations away... Uranium and natural gas will run out before 2075 – the UK was a net importer of natural gas on an annual basis in 2004 and 2005 and is expected to become a net importer of oil by about 2010. And while at the moment there is still plenty of coal, reserves will last no more than another 200-300 years. The idea that nuclear power is a solution to our desire for abundant, 'free' energy is a myth, thanks to problems associated with over-capacity and over-generation and waste disposal.

Only the use of renewable energy, in all its forms (sun, wind, water, wood and biomass), plus an increase in efficient use of the energy resources we have, will prevent us bringing the earth to its knees in order to satisfy our immediate needs. And it's not

Oil-: 2000 - 2045

Uranium-: 2000 - 2075

Natural gas-: 2000 - 2085

Reserves of energy at current rate of consumption[E]

Coal-: 2000 - 2700

Renewable energy-: 2000 - 3000 and beyond

only resource depletion that threatens our planet: the impact of the current output of CO_2 is causing problems that would become much much worse if the remaining reserves of fossil fuels were to be burned.

The UK Government's current strategy for ameliorating our impact on the environment consists of a number of schemes making up the Climate Change Programme, including the Energy Efficiency Commitment, the Renewables Obligation and the Climate Change Levy Exemption for Renewables. The Programme's aim is to reduce the production of greenhouse gases by 12.5 per cent and CO_2 emissions by 20 per cent by 2010. However, a number of independent institutions and campaigning organisations are calling for a direct carbon tax on all CO_2 emissions as the only way to achieve the targets set by the Kyoto Protocol. (For further details see Resources.)

In 2007, CAT submitted evidence to the Draft Climate Change Bill and a full report, outlining a 100 per cent renewable energy strategy for Britain. The report: *zerocarbonbritain*, is a 30-year update on CAT's original 1977 Alternative Energy Strategy. It is a standalone integrated energy policy for Britain; a practical blueprint for a sustainable future. (For further details see Resources.)

Energy efficiency and the 'negawatt'

Reduce the amount of energy you use for the same results – i.e. use a low energy bulb – fewer watts, same amount of light

Striking inequality, uncontrolled consumerism, increasing attacks on the environment, wastage of limited fossil resources... As far as energy is concerned, the current state of affairs is going from bad to worse. Nevertheless, we continue to produce and consume more energy than we actually need, and we conveniently bury our heads in the sand instead of thinking of the consequences. Future generations will regard us as greedy and thoughtless super-consumers who carelessly passed on a legacy of pollution to their children.

But is this inevitable? Can we put an end to such irresponsible behaviour without reducing our standard of living? There are many simple common sense solutions, which everyone can immediately put into practice. They are based on energy efficiency that reduces the quantity of energy needed for operating a single machine and making better use of energy in order to reach a better and lasting quality of life.

For example, a single action such as simply planning a house's orientation towards the sun, decreases its heating requirements and hence its energy consumption by 15-30 per cent.

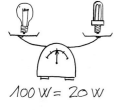

$100 W = 20 W$

Another example: replacing an ordinary 100W light bulb with a low energy lamp that consumes 20W, uses one fifth of the energy to produce the same brightness. The necessary amount of electric power has thus been reduced by 80W. In other words, the replacement of the original lamp produces '80W less'. This is called, for the purposes of this book, the 'production of negawatts', or the consumption of negative amounts of watts!

How to produce negawatts

In your house there will be real 'negawatt' spots

Becoming negawatt producers therefore means giving up our energy greedy habits for more energy conscious activities. It also means making the best possible use of energy, rather than simply continuing to consume more and more.[1] Far from going back to the candle or the oil lamp, our aim should be to track down wasted watts and eliminate them by using available energy more effectively and, wherever possible, looking to renewable energy sources.

[1] Amory Lovins, founder of the Rocky Mountain Institute came up with the notion of 'the production of negawatts', a concept similar to the production of the electric kilowatt. It was expressed in real economic terms: to sell the negawatt the Institute put to their clients the instant energy savings to be made by fitting a new heating system. In this book, the concept is applied to all forms of energy and not just to electricity.

Three 'negawatt' winners:

- **Consumers**: who see their energy costs decrease
- **The job market**: as new jobs are created via the installation of new and better equipment and the decentralisation of energy companies
- **The environment**: because the least polluting form of energy is that which isn't produced at all!

There are many potential negawatts hidden in every home. They hide in the insulation, or lack of it, in the heating system, in domestic hot water systems, lighting networks, and in most electrical appliances. You'll also find negawatts in the fabric of your home, around your windows, your taps, your television set, and even in the direction your house faces! *The Energy Saving House* encourages you to hunt them out using the 'negawatt' approach.

Information point

Energy: output

Output is the power produced or consumed per unit of time. The unit for measuring electrical output is the watt (W).
1000W = 1kilowatt (or 1kW)
The unit measuring energy is the Joule. Often the term used, particularly for electricity, is the kilowatt hour (kWh), which corresponds to the energy consumed by an appliance with an output of 1kW in 1 hour.
1000W in 1 hour = 1000Wh = 1kilowatt hour (or 1kWh).
So, a 100W bulb lit for 10 hours consumes 1kWh of energy, and a 500W halogen bulb will consume the same amount of energy in 2 hours.
It is also possible to express energy consumption in terms of the amount of energy that would be supplied by a tonne of petrol. This is also referred to as TPE tonnes of petrol equivalent. 1 TPE is equivalent to 11,600kWh.[1]

---

[1]International coefficient for conversion in all countries except France which has chosen a different coefficient for its energy comparability (1 TPE = 4500kWh) not taking into account the heat lost during electricity production in thermal and nuclear power stations.

References

(A) International Energy Agency statistics, 2000
(B) *IEA Energy Statistics* ©OECD/IEA, 2006, World: TPES/Population,
http://www.iea.org/Textbase/stats/index.asp
lettering, colour and line modification by CAT

(C) Unesco, 1984

(D) Observatoire de l'energie, 1999

(E) Conference mondiale de l'energie, 1989

2. **Planning**

Preparation is the key

Bioclimatic design

Build your house to work with the prevailing climate, not against it

Whether you're building a new house or renovating an old one, you'll maximise any energy savings by identifying the most significant areas for potential negawatts during the planning phase of any project. Apply a few simple principles: take account of the immediate environment, collect and store the sun's heat energy, insulate carefully, gain the maximum benefit from daylight, create shelter from wind and solar shading where necessary... It takes just a little common sense to build in harmony with the environment and the climate: this is why this approach to planning and design is called bioclimatic or passive solar design.

Following this pattern makes it possible to build a healthy and comfortable home that does not cost significantly more than a conventional house, has reduced heating costs, and no need for expenditure on air-conditioning (which is not usually necessary in the UK). Though set up costs may be higher for this type of design, the rate of return makes the initial outlay worthwhile.

Bioclimatic or passive solar house

Planning stages

The basics of bioclimatic architecture in eight steps

1. Study the ground you are building, or sited, on, the immediate environment and the local microclimate (sun, wind, vegetation).

2. If renovating an existing house, do a sketch plan of the present state of the house and re-arrange the function of the different rooms according to the orientation of their outside walls.

3. Insulate carefully to store heat in winter and to prevent overheating in summer.

4.Trap the sun during warm periods via glazing in the form of a conservatory or in high mass walls, both of which can also protect the body of the house from extreme heat in summer.

5. Where possible store energy in the mass of the building and reduce temperature variation with the help of thermal mass.

6. Limit unwanted infiltration of air in the form of draughts and encourage a natural renewal of air in the house by using either natural ventilation (windows, air bricks etc) or by using a controlled ventilation system.

7. Let as much daylight as possible into the house but beware of overheating.

8. Finally, choose an appropriate, low pollution heating system.

Planning bioclimatically means rediscovering the art of being at one with the environment: the rules of bioclimatic design are simple common sense, too often overlooked when the aim is to build houses fast, to a standard format and as cheaply as possible. Building or renovating a house according to bioclimatic methods requires a longer planning phase and is therefore a little more expensive. However, it is worth taking the trouble: a house is not an impersonal product for instant consumption, but a living space in which you should feel comfortable and in harmony with the natural environment.

'When I was working on the design for a solar building, I was intrigued by its thermal properties: the building itself acted as a heating system. The lounge was conceived as both a living space and to capture heat from the sun (as a solar collector). Its south facing windows both had the best view and let in the sun's heat in winter. I began to imagine how its thermal characteristics would affect the building's inhabitants. Its purity of purpose is similar to the warmth of the fireplace…and the coolness of an Islamic garden: the hearth is a warm dry refuge in a cold world and the oasis a place of fresh air and coolness in the heart of the desert.'

Lisa Heschong, *Architectures et volupte thermique,* editions Parentheses

The environment and your ecological footprint

The first, invaluable, step in planning a bioclimatic house is to analyse the environment around the site of the project. You need to know from which direction the prevailing wind blows, to look at the lie of the land and the local plant life and to determine if nearby buildings create shade at certain times of the day. Try and offer as little surface area to cold winds as possible, opting for orientation towards the sun, ensuring you collect the optimum amount of direct sunlight. The footprint, or compactness, of the building is also economically very important with regard to both energy and financial costs. Heat loss basically depends on the surface area of those walls that are in contact with the outside atmosphere: for the same volume of living space and the same floor area, a more compact/squarer dwelling consumes less energy.

Information point

Density coefficient c

Density coefficient c is the ratio S/V between the surface area of the outside walls of a house S and the volume of the living space V. The lower it is, the more compact the building. A coefficient of less than 0.70 is a very compact design..

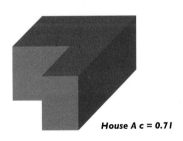

House A c = 0.71

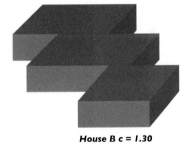

House B c = 1.30

Compare two projects, for example: house A has two storeys and house B only one, but both have the same amount of floor space (120m²) and the same volume (250m³). If the walls of the two houses were made from the same materials, you might think that their energy consumption would be identical. However, this is not the case because house B is far less compact: the total energy loss through the walls and floor is 82 per cent higher than that of house A. The energy requirements are thus much higher for a single storey building, even if volume and surface area are identical. Bioclimatic design does not specify extreme compactness, although it is important to know when designing a house that any decrease in the building's compactness automatically generates energy consumption and higher running costs.

Design your home to make the most of the sun

Any designer, architect or renovator must combine maximising the contribution of the sun's heat in winter with protection from it in summer and between seasons. As an indication, here are some guiding principles on which a number of bioclimatic houses are designed:

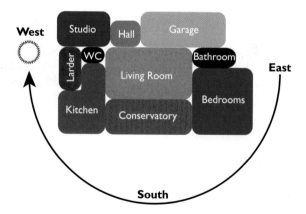

- The rooms that are mostly used during the day and evening should be oriented towards the south if at all possible.
- Bedrooms should be situated to the south and the east, where they will benefit from the early sun, and yet be cooler than west facing rooms at the end of the day.

- Limit the sunlight hitting south-west facing kitchen windows as this is often a cause of overheating (in practice overheating is not a significant problem in the UK).
- A conservatory or veranda on any side of the house creates an intermediary space between inside and outside, providing a heat source in winter if facing south. If large enough it could be used as an eating space accessible from the living room, kitchen and even the bedrooms.
- Spaces with little or no heating (porch, workshop, garage) are best sited facing west or north. If the prevailing winds are strong, an entrance porch will be necessary to prevent cold air entering the house.

The practicalities

Initially you should look to have an area of glazing corresponding to the surface of the floor in the following ratios:

Orientation	Ratio of window area to floor area
South	20-35%
East and west	10-25%
North	0-10%

These values are approximate. Beware of the danger from overheating if there is a surplus of glazing: it is important to protect against excessive heat from the sun in summer using plant screens, overhanging eaves, exterior blinds or pergolas.

Insulate with care!

Insulation is always beneficial: in winter it helps stop heat escaping, in summer it keeps the house cool by keeping heat out. Insulation also helps prevent condensation and the unpleasant sensation of 'cold walls', which often leads to overheating in order to create desired comfort levels. During the renovation of a house, insulation work on the roof space must take priority (on average, 25 per cent of all energy loss from a house occurs through the roof with a further 15 per cent lost via ventilation and infiltration

– draughts and air movement though cracks in the construction, 10 per cent through the windows, but the largest single part, 35 per cent, is lost through the walls.) When insulating, the first 100mm of insulation is always the most efficient. The optimal thickness depends on the climate. The recommended first layer of material, mineral wool for example, is 200-250mm for exterior walls, 300-400mm for the roof and 150-200mm for a floor above a cellar. In the UK, the recommended loft insulation is 270mm and although this figure rose in the most recent building regulation updates, we are still 30 years behind the standards of Denmark.

Heating demand drops rapidly with the increase in insulation in a house's walls[1]

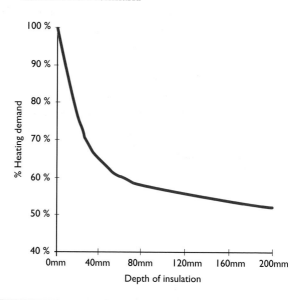

Thermal mass

Most of us are aware that a solid stone wall, sunny and well sheltered from the wind, remains warm long after the sun has gone down and we've all, at some time or another, been surprised by the refreshing chill of a cellar on a hot summer's day.

---

[1]GEFOSAT results: using PLEIADES & COMFIE software.

Heavyweight or dense materials have a dual role: heat storage in winter and absorption of excess heat in summer

The high thermal mass of the wall and the cellar is the cause of these phenomena: solar heat stored during the day in the mass of the stone wall is slowly diffused towards the exterior, because of this the sun's heat does not manage to raise the temperature of the cellar's walls, leaving the interior temperature close to the annual average.

To build with high thermal mass means using heavy materials for the interior walls of the house in order to store solar heat and to avoid fluctuations in internal temperature.

Conversely, a house with low thermal mass will heat up rapidly and (unless designed to do so in other ways) be unable to store solar heat. There will be marked differences in internal temperature and a greater risk of overheating.

Compare two identical houses, one with high thermal mass and external insulation, the other with low thermal mass and internal insulation… If the surface area of windows on the south front of the houses is gradually increased, the heating requirements for them will change as follows[2] :

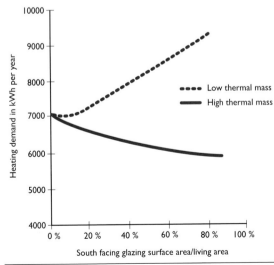

[2]Calculations: GEFOSAT, software: PLEIADES & COMFIE. Based on weather conditions for Montelimar. Based on a house 100m^2, 250m^3, internal wall insulation (low thermal mass) or exterior walls (high thermal mass), aluminium framed double glazing, east, west and north facing glazing: 1.6m^2, heating temperature 15°C at night and 19°C during the day.

In a house with low thermal mass an increase in the window surface area may increase the energy required to heat it, as the storage of energy cannot be carried out effectively. In a house with high thermal mass, on the other hand, the result is positive: the solar energy captured in the walls largely compensates for the losses from the increased window area, resulting in a decrease in the energy needed for heating. However, south facing double-glazing always has a net postive energy balance.

In winter, a bioclimatic house with high thermal mass benefits from the same phenomenon. When solar rays penetrate the windows, they hit the interior floor and walls: if these walls have high thermal mass (for example in a wall with insulation on the outside not within it), they slowly radiate their heat into the house. Conversely, in summer thermal mass allows a room to benefit from the chill of walls cooled down at night by ventilation. Building using thermal mass is a useful means of natural air-conditioning.

In both cases thermal mass plays a significant role in maintaining and controlling interior comfort levels by minimising variations in temperature.

Hints and tips

- Interior insulation, as well as suppressing thermal mass, exposes the exterior wall to constraints of the climate
- High thermal mass is most useful in a house that is occupied year round
- Avoid putting fitted carpet over a flagstone floor – it decreases the thermal (mass) properties of the floor

What should you choose?

Comparative tests on three French houses.

Let's compare three houses with the same surface area (100m²) and the same volume (250m³).

The first is a very traditional house, built to current standards, but without any particular thought having gone into the orientation and glazing of the building.

The second house resembles the first in all aspects and has exactly the same surface area of window glazing. However, it has been better sited with regards to its orientation and distribution of windows, the heating is turned down to 15°C at night as opposed to 19°C as in the first house. The shutters are used correctly in winter, that is, closed to keep the heat in. The third house has a surface area and volume identical to the first two, but utilises various components of bioclimatic architecture: a conservatory/veranda with access into the house, heavy supporting walls to the conservatory, good night ventilation in summer, insulating exterior blinds, and increased insulation to the walls.

The table shows the three houses' energy requirements for heating and air-conditioning given the same site:[3]

House number 2 is better managed and correctly oriented and ventilated, reducing its heating and air-conditioning needs by a third without any further expense!

	House 1 standard	House 2 well oriented	House 3 bioclimatic
Area	100m²	100m²	100m²
Volume	250m³	250m³	250m³
Winter temperature	19°C all day	19°C day 15°C night	19°C day 15°C night
Glazing	16m² 3.2m² of which is south facing	16m² 11.2m² of which is south facing	28m² 22m² of which is south facing
Winter nights	shutters open	shutters closed	shutters closed
Summer days	shutters open	shutters closed 85% of time	shutters closed 85% of time
Wall insulation	70mm interior	70mm interior	100mm exterior
Roof insulation	140mm	140mm	200mm
Heating and air-conditioning demand	14,300kWh	9420kWh -34%	5070kWh -65%

[3] Calculations: GEFOSAT, Software: PLEIADES & COMFIE. Based on 8760 hours for an average year's weather taken from readings in Rennes, Trappes and Montelimar.

House number 3, which is bioclimatic, has its thermal energy requirement in both winter and summer reduced by two-thirds! Air-conditioning is not required, even in very sunny areas of the house. To achieve results like this you don't need to spend huge amounts, but planning for energy conservation from the very beginning of the project is essential. If you're building a house to last you a lifetime it's surely worth a few weeks' extra thought at the design stage? In the UK careful shading and cross ventilation may be all that you need in most situations – inexpensive if included in the design at planning stage.

Energy conservation, comfort and health

Energy saving, thermal comfort and good health are inextricably linked

Designing a house to be healthy and comfortable doesn't mean it has to use a lot of energy, rather the opposite. If the walls of a house are badly insulated, it results in significant heat loss through those walls, leaving the occupants no option but to raise the air temperature to a level they find comfortable, no matter how uneconomic or wasteful of energy.

In effect the temperature felt by a building's occupants is not the air temperature, it is an average taken from the air temperature and the temperature of the enclosing walls. This explains why you might feel cold in a place that is badly insulated, with cold walls, even though the air is at a comfortable temperature. For example, if the temperature of the air is 19°C but that of the walls is only 15°C, what you sense will be 17°C. You would have to boost the air temperature as far as 24°C in order to sense your surrounding temperature as 20°C and feel warmer.

Hence, in winter, the occupant of a badly insulated house will constantly overheat the house in order to be sufficiently comfortable. And remember each degree Centigrade of additional heating consumes 7–10 per cent more energy! In summer, the occupants of a house that lets heat in too readily, could be tempted to install air-conditioning: this will result in an increase in the consumption of electricity, and perhaps even in a health risk if the air-conditioning is set too high or if the system is badly maintained.

Air velocity is also a significant parameter of thermal comfort: it must be lower than 0.15 meters a second if it is not to be felt as a cold draught. Otherwise, a feeling of cold may set in and you may even risk catching a chill. If air velocity rises from 0.10 to 0.30 m/s, a drop in temperature of 3°C is felt! (This is known as 'wind chill'.)

The relative humidity of the air should be between 40 and 65 per cent for optimum human health. Insufficient humidity dries the respiratory mucous membranes, which then stop filtering dust and pathogenic germs. Too much humidity upsets the body's ability to regulate temperature because evaporation from the skin stops, which in turn increases perspiration as the body attempts to prevent overheating.

A house in which the temperature is too low gets humid very quickly. There is a high risk of condensation on the walls, which can encourage the development of mould, of microbial germs, and of acarina (dust mites) that can cause respiratory disorders (such as asthma – particularly in children). Too much humidity also contributes to the deterioration of the interior of houses (grey spots, saltpetre, damage to fabrics and papers).

Other factors must also be taken into account: for regular renewal of air, ventilation must be about 0.5 to 1 times the volume of the house per hour or 25m^3 per hour per person. Too much ventilation increases heat loss to the outside, whereas too little encourages humidity.

Poor combustion within a heating system will affect the quality of the air by releasing CO_2 and CO into the house. It is important that all appliances/fires have their own source of fresh air nearby. Finally, a lack of natural light in a house often leads to the use of artificial light, which increases the consumption of electricity and can also lead to visual tiredness.

Good health, comfort and low energy consumption are all inextricably linked to the temperature, humidity, air speed, air renewal and interior light levels of the house you live in. Thermal comfort is not an expensive luxury, regulating it makes it possible to reduce your energy bills and live a healthier life.

Don't go it alone...

So, now you want to build or renovate minimising your eventual energy consumption and the effects of your project on the environment. Where do you start? If you are building a house for your own family, you can do both the design and building yourself, you can manage the project but employ builders and architects who will do the work, or you can employ a company to deal with all aspects of the project and merely hand over the keys to you on completion. However, the less of a hands-on role you take, the less likely it is that energy efficiency and environmental concerns will be fully taken into account.

Although it is generally not obligatory, the help of an architect is advisable for most projects. Still, he/she should be carefully chosen... If your chosen architect doesn't specialise in bioclimatic design, it would be helpful to employ a thermal engineering and design company to work alongside them: there are plenty of completed projects which prove that the additional costs of an architect and an engineering and design company are covered by the long term savings earned by the application of their expertise.

3. Materials
Choosing materials

Materials and energy: criteria for making the right choice

The type of material used for walls, floors, and windows directly influences the energy consumption in a 'negawatt' house. The insulating quality of the materials is crucial, but other criteria also play a role: thermal inertia in order to better balance variations in temperature, acoustic performance for better sound comfort, and hygroscopic quality which allows the house to breathe and to let out excess humidity. Lastly, it is important to check the environmental credentials of the materials used: is a lot of energy required to produce them? Are any natural resources or the environment adversely affected by their production? Can they be recycled? Are they a source of toxic emissions?

Taking into account the above criteria, this chapter examines the advantages and disadvantages of the principal building materials used in a house from floor to roof.

The range of insulators

Information point

Sound insulation between rooms

Contrary to what one might assume, thermal insulation does not necessarily insulate against sound.
There are effectively two solutions for sound insulation between two rooms: constructing a wall using dense, heavyweight materials, or putting a fibrous material between two lighter weight partitions. And if this material is flexible and springy (like glass wool, hemp or linen, cocoa fibre or sheep's wool), the better the sound insulation will be. In contrast, very compact materials such as polystyrene or polyurethane are very poor sound insulators[1]. For two rooms one above the other, a floating floor (not anchored at the walls) on top of a non-compacted cork insulation, cellulose panels, or rockwool, offers the best sound insulation.
Do call in a specialist: acoustics is a complex and difficult subject area within which it isn't easy to improvise and where mistakes can be costly to rectify.

[1] There is an 'acoustic' polystrene available, but its performance is inferior to fibrous sound insulators.

Any material's capacity to insulate rests not on the nature of the material itself, but on the amount of air trapped within it. The drier and the more motionless that air, the less heat can get into or past the material, the higher its insulating capacity will be.

The less dense insulators can be split into three main groups:

- Mineral based insulations: fibreglass, rock wool, vermiculite, perlite, cellular glass or expanded clay
- Insulators containing aerated plastic: expanded polystyrene, extruded polystyrene and polyurethane
- Organic insulation: cork, wood fibres, cellulose, hemp, flax, cotton, coconut fibre, wool. These perform well, they can be recycled and producing them requires only a small amount of energy.

A heavy, very dense material, such as an old stone wall, seldom insulates well. Although it may be useful in summer because it has high thermal mass, it will readily lose heat in winter. Walls with integral insulation properties[1] such as honeycombed brick, aerated concrete, wood twists, and hemp concrete offer a good compromise between insulation and thermal mass.

U-value is a measure of the rate at which heat is transferred through a material or building element. UK Building Regulations or your local/national certification board will stipulate minimum U-values or equivalent for different building elements (roof, walls, floors). In the UK the process of certifying that a material is 'fit for purpose' as an insulator is covered by British Standards of BBA (British Board of Agrement). Methods for calculating the U-values required for your house and the materials used to construct it are given on pages 20 and 21 of the 2006 UK Building Regulations Approved Document L1 (see Resources).

[1] Also termed divided insulation or 'single' wall

Depth of various insulation materials needed to equal the insulation value of a 90mm concrete wall

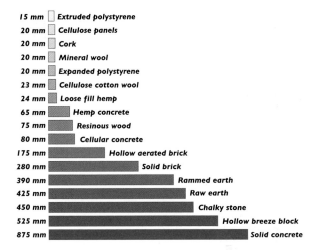

15 mm	Extruded polystyrene
20 mm	Cellulose panels
20 mm	Cork
20 mm	Mineral wool
20 mm	Expanded polystyrene
23 mm	Cellulose cotton wool
24 mm	Loose fill hemp
65 mm	Hemp concrete
75 mm	Resinous wood
80 mm	Cellular concrete
175 mm	Hollow aerated brick
280 mm	Solid brick
390 mm	Rammed earth
425 mm	Raw earth
450 mm	Chalky stone
525 mm	Hollow breeze block
875 mm	Solid concrete

The practicalities

How to decide on the level of insulation you need

The means by which insulation conducts heat is called thermal conductivity. It is expressed by the coefficient lambda: which refers to the amount of heat a material can transfer from one side to another and therefore good insulators have a low figure. The insulating quality of a wall of a certain thickness is measured in terms of its thermal resistance (strength) R defined as follows: R=thickness/lambda. A correctly insulated wall will therefore consist of a sufficient thickness of a material with a low coefficient of lambda conductivity.

In the UK the insulation value of a construction material is summed up by its U-value. U-value is the thermal transmittance of a material and is expressed in units of watts per square metre per degree of temperature difference (W/m²K)

Don't worry about putting in insulation of a greater value[1] than that recommended: adding several extra centimetres of insulation is only a little more expensive if you do this at the outset, whereas it is much more expensive to install extra or reinstall after a building is finished.

Recommended U-values (W/m²K) for construction elements

	U-value
Pitched roof with insulation between rafters[2]	0.2
Pitched roof: integral insulation	0.2
Pitched roof: insulation between joists	0.16
Flat roof[3]	0.2
Walls, inc. basement walls	0.3
Floors, inc. ground floors, basement floors	0.22
Windows, doors and rooflights[4]: glazing in metal frames	2.2
Windows doors and rooflights[4]: glazing in wood or PVC frames[5]	2.0

Notes
[1] e.g. to avoid the heat at the end of the day ending up crossing exposed walls, in summer.
[2] Any part of a roof having a pitch of 70° or more can be considered as a wall
[3] Roof pitch not exceeding 10°
[4] Rooflights include roof windows
[5] The higher U-value for metal-framed windows allows for additional solar gain due to the greater glazed proportion

Floor insulation

Installing flooring in the ground floor of a house can be approached in two ways – flooring laid directly on the ground (e.g. concrete slab on solid ground) or suspended flooring (ventilated basement).

Putting the floor straight onto solid ground allows you to make use of the thermal mass of the earth in addition to that of the house.

With a heated floor, the best solution is to insulate beneath it.

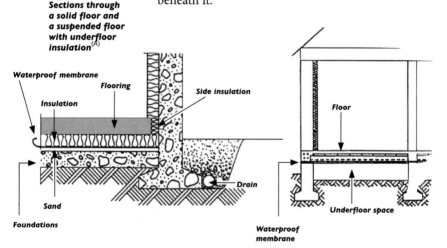

Sections through a solid floor and a suspended floor with underfloor insulation[A]

Waterproof membrane

Flooring

Insulation

Side insulation

Floor

Drain

Sand

Foundations

Underfloor space

Waterproof membrane

If the main walls are insulated on the inside it will be necessary to allow for peripheral insulation fitted at right angles to the floor to reduce the risk of draughts between the floor and the ground outside.

There are three additional solutions:
- lengthening the insulation on the internal face of the foundations
- installing peripheral insulation horizontal to the floor, for 1m width beneath it
- fitting horizontal insulation beneath the whole surface of the floor before putting down the flooring. This is the most common method in the UK currently.

This last option can increase the insulation value of the floor by up to 40 per cent over the first two options. It is, however, a little less useful in summer because the heat within the house will find it more difficult to escape via the ground.

Floors can also be insulated by replacing the gravel in the concrete mix on which they are set with synthetic aggregates or by supporting the slab on inert insulation (cork, foamed glass, perlite, expanded clay beads).

Supporting walls

Whether they are made out of wood, concrete, earth or stone, the choice of the principal component of a house's walls is not without consequences for its energy quality and thermal comfort.

Wood has a lot of advantages as a house building material: it is natural, healthy, recyclable and needs only very little energy to produce and put in place. A wooden framework with an insulating filling, however, offers little in the way of thermal mass. This disadvantage can be compensated for by the installation of solid flooring and heavy interior walls[2].

Precast aerated concrete blocks are largely used in the self-build construction industry. A wall made of solid concrete blocks of the same volume must be very well insulated in comparision. The low porosity of concrete makes it a bad humidity regulator. In addition, solid concrete has a rather average thermal mass quotient, lower than that of a wall of blocks or formed concrete.

---

[2]Interior walls can be concrete, brick or raw earth. One recent technique improves the mass of wooden walls by covering their slats in a mixture of hemp, lime and plaster. This method also has the advantage of protecting the wood against rot and allowing the walls to breathe regulating humidity and temperature. However, they will still have a relatively poor U-value.

Aerated concrete is formed from a mixture of sand, cement and lime which, together with aluminium powder (1 per cent), creates a multitude of air bubbles. This substance has good thermal, acoustic and hygroscopic qualities. It has excellent fire resistance properties. A 300mm block has an average thermal mass rating and the Centre for Alternative Technology (CAT) would recommend using complementary insulation. Its low mechanical resistance means, however, that great care needs to be taken during installation.

Terracotta bricks are made from argillaceous earth. Traditional bricks are generally more insulating than concrete imitations but, like them, they require either interior or exterior insulation. Aerated bricks of significant thickness, however, make it possible to construct a wall that is at the same time carrier and insulator. In addition to its thermal qualities, terracotta is a hygrometric regulator.

Raw earth is a traditional building material, which has excellent hygroscopic qualities and high thermal mass. It has the advantage that it is almost universally available, needs very little energy in preparation and can be recycled[3]. But walls made out of raw earth alone do not offer adequate insulation. They are best used for interior walls or at the rear of a conservatory where they act as conductors for solar radiation.

Stone is not often used in new buildings nowadays. Its thermal and environmental qualities are similar to those of earth, but it doesn't have the same hygrometric qualities.

[3] This is used in the form of pisé (rammed earth) (earth with added lime packed down between two pieces of shuttering) or in blocks of adobe (in this case bricks of raw earth with added chalk – approx 5 per cent – are pressed into blocks with a block maker and then dried).

Roof insulation

Precautions to take if you are using fibre-based insulation.

Uncovered, fibrous insulation materials loose or in rolls can release fibres into the atmosphere that can be damaging to health – the effects are still not fully known. When installing such insulation, always wear a mask and limit the use of such materials to within partitions and ensure that any areas adjacent to living accommodation are completely airtight. Once installed there should be no passage of fibres to the living space. Fibrous insulation which has been squashed or crushed or which has slumped loses its efficacy: when the material is laid in a loft, you must ensure that there are air spaces for ventilation. Lastly, rodents love to work their way in and out of insulation wools. To prevent their nightly ramblings, you'll need to seal insulated false ceilings completely.

In the case of an unoccupied attic, insulation can be put directly above the ceiling in batts or in rolls. The best method is to lay it out in two crossed layers to avoid any heat loss in the junction between the two rolls.

In the case of flat roofs, the insulating material should be placed under or above the waterproofing layer. Several techniques are used for sloping roofs.

Installing panels or rolls of either mineral wool or quilted sheep's wool is a relatively inexpensive solution, although more time-consuming. The best method is to lay out the insulation in two crossed layers and to secure it using fixed pegs. Metal or wooden rails fixed along these pegs are used to support the interior surface. Installing bulk insulation between the rafters means that a significant depth of rafter is needed plus something to affix the insulating material to so that it does not slip. This does not allow for continuous insulation and leaves many thermal bridges around rafters.

Prefabricated insulating slabs combine the following three elements in one construction panel: the ceiling (made from plaster or chipboard), heat insulation and cover[4]. Plastic foam bonded to plasterboard is common in the UK. Though the cost of this type of component is higher than that of more traditional materials, it is a lot easier to install.

Roof insulation: the interior facing and then the insulator are placed across the rafters in one or two layers.

---

[4] Damp-proofed facing which can be put against lathing

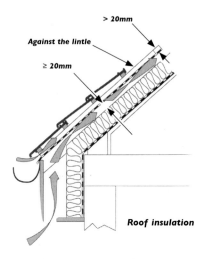

> 20mm

Against the lintle

≥ 20mm

Roof insulation

Hints and tips

- With organic or mineral wool insulation put a vapour barrier on the inside face of the insulation
- Include ventilation at the top and bottom of the covering and an air gap of at least 20mm between the insulation and the roof battens to allow for ventilation
- In very sunny areas don't worry about over insulating the roof – it will prevent heat penetration at the end of the day

The practicalities:

Price and performance of popular insulation materials

Comparison between different insulating materials

	Thermal conductivity in W/mK	Price not including fitting in £/m²
Polyurethane	0.025	7.50 - 15.00
Extruded polystyrene	0.029	10.00 - 13.00*
Cork	0.036	10.00 - 15.00
Loose fill paper - Warmcel	0.036	10.00 - 15.00
Flax wool	0.038	12.00 -13.00
Sheep's wool batts	0.039	8.00 - 10.00
Rockwool	0.04	1.00 - 5.00**
Standard glass wool	0.04	1.00
Hemp wool	0.04	no longer in general use in the UK
Loose fill hemp	0.048	3.50 - 5.00
Wood wool cement board	0.075	10.00 - 17.00***
Hemp and lime mix	0.12	not in general use in UK

0

* 50mm is usual (wall £10.00/flat roof £13.00)
** £5.00 wall insulation
*** £9.50/m² or £17.00 1200 x 500mm
    50mm thick   100mm thick

Insulation and humidity

Certain materials naturally regulate ambient humidity

Water sensitivity of insulation.

If it takes the place of air in the spaces in any insulating material, water decreases its insulating properties hugely: water has in effect 28 times the heat conduction potential of air. Insulating materials with aerated structures – such as cork, polystyrene, polyurethane or cellular glass are thus not very sensitive to water. On the contrary, fibrous insulation (like cellulose panels, mineral and plant based wools) is prone to water infiltration which greatly diminishes its insulation potential and encourages moulds. With this type of insulation it is advisable to fit a vapour barrier to the interior facing (adjacent to the living space) in order to avoid condensation appearing in the insulation material.

A family of four people produces between 10 and 20kg of water per day in the form of vapour caused by breathing, cooking, using the shower and bath. A healthy house should maintain an interior environment that is neither too dry, nor too wet.

However, the majority of modern insulators based on hydrocarbons (such as polystyrenes and polyurethane), or the plastic coatings on ceilings, are impermeable to water vapour. They create a water-tight barrier in the wall.

Certain materials (such as aerated bricks, aerated concrete or hemp concrete) are, on the contrary, good hygroscopic regulators. Thanks to their density they can provide insulation or thermal mass.

To help prevent the development of micro-organisms and mould, you should ventilate the house well, constantly renewing the air. However, over-ventilation can waste energy. Porous walls are a potential solution, such as those made out of earth or hollow bricks. Their porosity enables them to absorb the excess water vapour and then to restore it if the air is dry. They help regulate humidity naturally[5]. Remember, though, that you still need proper ventilation even with hygroscopic materials.

A cement based plaster coat lets through 45 to 50 grams of water vapour per m² per day, a plastic coating 50 to 60 g/m²/day, and a natural lime plaster 300g/m²/day.

---

[5] Nevertheless it is necessary to ensure that the interior and exterior finishings (such as breathable paint or limewash) have greater permeability than the wall itself in order to avoid creating a watertight barrier.

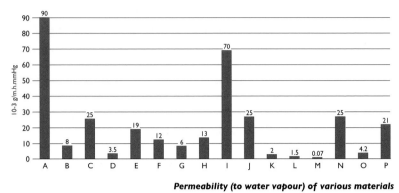

Permeability (to water vapour) of various materials

A	**Air**	**I**	**Mineral wool**
B	**Brick**	**J**	**Fibreglass panels**
C	**Hollow brick**	**K**	**Expanded polystyrene**
D	**Solid concrete**	**L**	**Polyurethane**
E	**Cellular concrete**	**M**	**Cellular glass**
F	**Lightweight concrete**	**N**	**Wood panelling**
G	**Coating mortar**	**O**	**Watertight plastic coating**
H	**Plaster**	**P**	**Lime coating**

Windows, glazing and joinery

Glass is an extraordinary material, which can now be used without fear of too much heat loss, but which is the right type of glazing to choose for a 'negawatt' house? The best option currently available is low emissivity, argon-filled (or gas-filled) double glazing.

Compared to single glazing, double glazing decreases heat loss by 40 per cent. It stops condensation on the inner window pane and decreases the unpleasant effects of cold walls. This effect can be felt from 2 metres away with single glazing, whereas with double glazing it disappears after just 1 metre.

Low emissivity glazing has a special interior coating that traps the infra-red radiation emitted by the walls and furniture. Its insulating capacity is reinforced, increasing the performance of the double glazing by 20 to 30 per cent for a cost of approximately 15 per cent more. With this type of glazing, the effect of cold

Information point

There are a number of certification schemes for construction materials and fittings accredited by UKAS (see Resources).

walls disappears after just 600mm. Low emissivity glazing is already the standard in the UK, regulatory U-values cannot be met by double-glazing without it. The latest change has been from 'hard coated low-E' such as 'Pilkington K' to 'soft coated low-E' such as I-plus, available from the Green Building Store.

Which type of frames should you use? PVC is the best insulator, then wood (10 per cent less compared to PVC), aluminium with flexible seals (35 per cent), then traditional aluminium or steel (55 per cent).

PVC frames, however, are generally wider than other standard window frames, which decreases the amount of light they let in. They require relatively little maintenance but their chemical composition makes them difficult to recycle and high in embodied energy costs. Wood, the most natural material for any form of joinery, requires regular maintenance but well made, well maintained timber external joinery should last at least 50 years, if hardwood perhaps longer. Make sure that you do not use imported hardwood but wood from sustainably managed European forests. Native woods such as oak are well suited to use in window frames.

Aluminium frames are narrower, allowing more light in. They are relatively indestructible, but huge amounts of energy are required to produce them and their production is a cause of environmental pollution. If you choose this type of joinery, choose frames with flexible seals or fittings: this additional insulation stops condensation, a frequent problem on traditional aluminium windows, and it avoids the 'radiator effect' that happens if frames are exposed to the sun in summer. These frames also make it possible to reduce heat loss by approximately 25 per cent for an initial set up cost of 15 to 20 per cent more.

Average U-Values	W/m²K
Single glazing	5.4
Double glazing with 12mm air gap	2.6
Double glazing with hard low-e coating	1.8
Double glazing with hard low-e and argon gas	1.5
Double glazing with soft low-e and argon gas	1.2
Triple glazing	2.2
Triple glazing with argon fill	<1.0

Source: Green Building Bible, Vol 1, 3rd edition

In the UK, typical window configurations used are different[1]. The optimum air space is 20mm, and 4/20/4 is standard in most uPVC windows. 4/16/4 is very common in timber framed windows.

Another interesting development is timber frames with a layer of insulation incorporated – Vogrun from Ecomerchant for example.

[1]Double glazing is classified by three figures corresponding to the glass used and the width of the gap between the two sheets of glass in the unit. E.g. 4/10/10 glazing is made up of two sheets of glass, one of 4mm and one of 10mm separated by a gap of 10mm.

Hints and tips

- Close shutters and/or curtains in the evening: this will prevent heat escaping and lessen the impact of any cold coming from the window itself.
- Less than airtight joinery is the most common weak point in any window. Fitting flexible joints (metal or rubber) on the edges of a window can reduce your heating costs by 15 per cent.

Health and choice of materials

Current building regulations still say very little about the impact materials have on the environment or their possible impact on human health. Considering the possible risks to public health, it takes far too long to implement regulations in these areas. For example, the dangers to health from asbestos were known about for years before its prohibition (in France as recently as January 1997[c]).

- The components of substances like concrete, brick and some mineral wools mean that the factories that produce them are likely to emit a radioactive gas, radon. No study of public health currently makes it possible to measure the potential consequences of these substances on health.
- The accidental combustion of polystyrenes or polyurethane releases poisonous gases.
- Mineral fibres (rockwool, glass wools etc), when combined with asbestos, are classified 2B by the CIRC[6]: they can be carginogenic. As a precaution, these fibres should be used only in spaces that have no contact with the interior air of the house. This does not, however, eliminate the risks incurred during their manufacture, installation or removal. If at all possible avoid their use completely.

---

[6]International Centre for Cancer Research.

The embodied energy of materials

All materials consume energy in production, transport to the building site and installation. However, it is important to weigh the embodied energy of any material[D] against the savings in energy consumption it will give once the building is in use[7].

When designing a house with low fuel/energy consumption in mind, it makes sense to ensure that the amount of energy used to build it is not unreasonably high. Evaluating the embodied energy of a product can help you choose between various solutions: for example, per kilo of materials consumed, putting aluminium window frames into your house uses 50 times more energy than fitting the wooden equivalent.

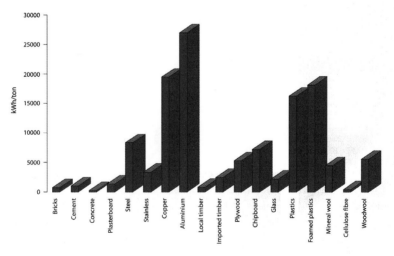

References
[A] Technical building journal
[B] *Guide de la thermique dans l'habitat neuf,* Editions du Moniteur, 1992
[C] Order no. 96-1133, 24 Décembre 1996
[D] Principal scource: *Logements à faibles besoins en énergie,* Oliver Sidler, 1997

---

[7] These values are not precise (they are very difficult to assess), but are in order of importance.

Which heat source should I choose?

What should the temperature in a house or flat be?

Current regulations recommend a temperature of 19°C (21°C for the very young and the elderly). This is fine for a living room (lounge, kitchen), but far too high for a bedroom where 15 or 17°C is sufficient and, in fact, better for your health: high air temperature dries out the nasal passages (a cause of snoring!) and can prevent a good night's sleep.

In a room where you spend most of the time sitting still, it is vital to have well insulated walls to avoid the disagreeable sensation of chilled air coming off the walls. Lastly, to avoid unwanted draughts you must make sure the difference in temperature between the floor, walls and the surrounding air doesn't exceed 4 or 5°C.

In summer, a temperature of 28 to 30°C is the maximum that is comfortable. Temperatures above that will mean introducing ventilation to cool the room by a few degrees.

Electricity, oil, gas or wood? Water or air storage systems? Individual heaters or central heating? Radiators or underfloor heating? Choosing the right heating system is never easy. You will have to take into account the installation and maintenance costs, the space you want to heat, the effects of your system on the wider environment and also the comfort level you are looking for. All too often the initial outlay is the only deciding factor. When it comes to buying heating systems, appearances can be deceptive. The cheapest systems to buy are often also the most expensive to run, the least effective and the most polluting.

Information point

Why is a poorly heated room often damp?

Air always contains a certain amount of water vapour limited by temperature and humidity. But there is a limit to the amount of water vapour that it can hold: for a given temperature there is a maximum amount of water that can be absorbed by the atmosphere. When this maximum is reached the atmosphere is saturated and the humidity is 100 per cent.

The colder the air temperature, the less water it can hold in the form of vapour: thus air at 20°C and 60 per cent humidity will, if the temperature falls to 15°C, feel damper and have an 85 per cent relative humidity level.

Relative humidity rises as the temperature drops and the risk of condensation on cold surfaces is increased.

This phenomenon explains why water condenses more readily on single glazing than on double glazing and is the reason for the unpleasant damp sensation felt in many poorly heated and badly insulated houses.

Hints and tips

- Put thermometers around the house and get to know what the temperature in each room really is. They should be fixed between 1.2m and 1.8m from the floor and out of direct sunlight.
- Any heater should be given the room to do its job properly: don't shade it or obscure it with a net curtain, piece of furniture or shelf.
- Pull the curtains at night: this will help keep heat from escaping through the windows.

Heating types

Information point

Heated floors and ceilings have been around for some years now, heated by electric cables sunk into the flooring or ceiling itself. With this type of space heating system it isn't possible to change to another energy source after its installation, and it is rather expensive in terms of energy and cost. Underfloor heating systems which work by piping hot water through the floor, can however, make use of any type of energy source to heat that water, and since they require a lower temperature than say radiators, they reduce energy consumption.

Individual heaters are generally designed to heat a single room. Central heating systems consist of a boiler that supplies hot water to a number of radiators, which in turn heat different rooms in the house. This has a number of advantages: good comfort levels, flexibility and ease of use, and the ability it gives you to choose from many types of energy source and to change the energy source used at any time. Boilers and radiators with vastly improved efficiency ratings have been developed over the past decade. The water that runs through a typical system does not need to be heated to as high a temperature as it used to be (50-60°C instead of 60-80°C). The heat that is emitted is gentler, more uniform, than the early central heating systems could provide and modern boilers are extremely efficient, especially condensing boilers.

Steel or aluminium radiators can be quickly readjusted and are best used in well-insulated houses and in rooms which are intermittently occupied. They are, however, liable to corrode and should have an anti-corrosion inhibitor built into or added to the system (as should all central heating systems) and they can sometimes be noisy.

Underfloor heating has also been improved since it was first developed nearly 30 years ago. It has little in common with the systems used in the 1960s, which were often too hot for comfort and poorly controlled. Today, efficiently installed, it should run at 35 degrees Celsius, compared to radiators which run at 75 degrees.

With this system, the whole floor area acts as a low temperature radiator: the entire floor is uniformly warm. Because of this and because such systems do not generally require additional ventilation, underfloor heating offers the greatest thermal comfort.

In an underfloor heating system warm water circulates in plastic tubes sunk into the floor or beneath the flagstones

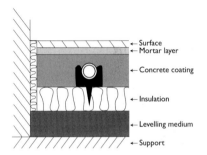

← Surface
← Mortar layer

← Concrete coating

← Insulation

← Levelling medium

← Support

The practicalities

Pros and cons of non-electric heaters

Type	Pros	Cons
Bottled gas heater	Radiant heat. Portable.	Possible risk of carbon monoxide poisoning if not correctly maintained. Releases a lot of water vapour.
Mains gas convection heater	Cheap running costs. Suitable for smaller spaces e.g. flats and apartments.	Requires flue through an outside wall. Requires annual service. Poor control. Low efficiency. Intermittent heating leaves mass of building cold.
Wood stove	Cheap to run. Uses renewable energy source. Pleasant association with traditional wood fire. Zero CO_2. Non-polluting if wood burnt dry and at high temperatures.	Regular maintenance important. Requires regular refuelling and annual service.
Paraffin stove	Pleasant radiant heat. Portable.	Risk of possible carbon monoxide poisoning. Releases water vapour. Expensive to run. Poor heat control. Intermittent heat means mass of building not warmed.

The practicalities
Pros and cons of electric heaters

Type	Pros	Cons
Convection heater	The cheapest to buy	Requires good insulation to keep running costs from rising. Dries the air and circulates dust. Layers warm air rather than heating entire space.
Radiant heater/ infra-red heater	Gentler, more even heat than convection heater.	High running costs. Around a 10 per cent energy saving on convection heating, but thanks only to its slightly more diffuse heat. Two or three times more expensive to buy than a traditional convection heater. Infra-red heaters are not recommended for use with children or in the vicinity of flammable materials.
Oil-filled electric radiator	Gentler and more even heat than convection heater. High inertia means stable temperature. Portable and well suited to use in poorly insulated rooms or those which are only used occasionally.	Running costs when used sporadically are equivalent to those of a radiant heater. Approximately one and a half times more expensive than conventional convection heater.
Storage heater	Runs at a constant temperature. Heats up fast.	Approx twice as expensive as a convection heater.
Fan heater	Heats up fast.	Dries air and burns dust. Causes a lot of air movement. Increases layering of warm air (warm air has tendency to rise leaving cold air beneath it).

Simple to maintain and very flexible, these devices are not however non-pollutants – they could indirectly generate nuclear waste and CO_2 emissions depending on how the electricity that powers them is produced.

The real cost of the kWh used

Gallons of oil, piles of wood, cubic metres of natural gas, electric kWh... What's their real cost?

A rigorous approach would compare the price of the energy to the kW hours used, that is the effective supply of the energy used in the rooms of the house. To do so would mean taking into account the true

The practicalities

Pros and cons of various central heating systems

Type	Pros	Cons
Water filled radiators	Well distributed warmth. Can run on a cheap form of energy intially and change to another at a later date.	High installation cost. Requires maintenance.
Underfloor heating/ceiling space heating via water filled pipe.	Can use cheap energy source. Comfortable heat. Non-drying heat, doesn't encourage dust. Not visible.	Approx 10-15 per cent more expensive to install than other forms of central heating.
Direct solar underfloor heating	Non-polluting inexhaustible resource that isn't affected by changes in the cost of conventional energy sources. Up to 40-70 per cent saving on space and water heating. Warm underfoot.	Expensive to install with a long payback period. Needs site suitable for solar collectors and may need another fuel source as a back-up. Not normally suitable in UK climate (requiring an interseasonal heat store and a very well insulated building).
Geothermal underfloor heating (uses a heat pump)	Potential for cutting cost of electric heating by 2.5 – 3 times. Warm underfoot. Can also be used to cool house when necessary.	Uses a coolant that can be harmful to the environment (they can also be designed to run on propane). Good house design could eliminate the need for a cooling system. Direct systems (with refrigeration liquid throughout the circuit) don't allow a change of heat source.

cost of the kW hours (billed by the supplier), and also the overall cost of the heating apparatus, its use, and its maintenance.[1]

The average UK household at the time of publication of this book consumes 21,816kWh of energy, so making sure you use energy in the most efficient way is vital. The chart on page 40 is a guide to comparative UK domestic fuel costs.

Electricity is still the most expensive form of energy. Choosing double tariff (split peak and off-peak) does not significantly change the energy cost per kWh.

---

[1]In the same way, the price of petrol at the pumps isn't the only criteria for choosing a car: the engine efficiency (mpg) and engine oil expenditure are equally important when calculating your car's running expenses in £ per mile.

John Willoughby's domestic fuel price guide

Fuel	Price (£)	p/kWh	£/GJ	Quarterly stand chg	Relative to gas	Rank	Kg CO$_2$/ kWh
Gas	4.02 p/kWh"	4.02	11.18				
	2.05 p/kWh" "	2.05	5.70	£21.57++	1.00	2	0.19
Electricity	15.03 p/kWh**	15.03	41.78	£7.13++	7.33		
(on peak)	11.11 p/kWh***	11.11	30.98		5.42	11	0.42
Electricity	20.98 p/kWh**	20.98	58.32	£17.16++	10.23		
(Economy 7)	11.55 p/kWh***	11.55	32.11		5.63	12	0.42
Night rate	4.41 p/kWh	4.41	12.26		2.15	9	0.42
Oil (35 sec)	36.5 p/litre*	3.47	9.64		1.69	7	0.27
Oil (28 sec)	31.18 p/litre*	3.24	9.01		1.58	6	0.27
Coal	175.00/tonne+	2.10	5.84		1.08	3	0.32
Anthracite	201.00/tonne+	2.21	6.15		1.08	4	0.32
LPG	35.34 p/litre*	4.95	13.75	£13.65	2.41	10	0.23
Wood	220.5/tonne•	4.17	11.61		2.04		0.03
Pellets						8	
Wood	147.86/tonne	2.80	7.78		1.37		0.03
Pellets						5	
Logs (B'leaf)	65.00/load••	1.97	5.47		0.96	1	0.0

"   based on B Gas DD price for first 12kWh/day
" " based on B Gas DD price for over 12kWh/day
**   based on nPower first 182kWh/q
*** based on nPower for over 182kWh/q
     (£20/yr discount available for DD)
*   based on 1000 litre delivery
+   based on 1 tonne delivery
•   15kg bags. 1 tonne + delivery (£49+VAT)
     bulk delivery (10 tonnes 80 miles)
••   cost in Lydney. Staking ratio 0.56, 9 GJ/m^3
All prices include VAT at 5%
CO$_2$ figures from SAP 2005

Prices obtained from nPower, British Gas, Welsh Biofuels, RHP and ' best practice' from local suppliers. Thanks to Alan Clarke for logs.

Price relative to gas now distorted by two tier tariffs and no standing charges.
++ if consumption over first tier you can use 2nd tier price plus equivalent standing charge.

Comments   Since April 06 – Electricity up 5% E7 up 5%
           Gas down 8%
           Oil 35s down 12%, 28s down 15%
           Coal no change, Anthracite up 6%
           LPG Previous prices (before April 2004) turned out to be 'introductory offers'.
           This is a more realistic price from Salkent up 10.5%
           Wood pellet – bagged down 9%
           Logs – no change

LPG or propane gas is cleaner than oil but its price depends very much on the current economy. However it can be a reasonable alternative system if you intend to switch to natural gas at a later date[2]. Oil for heating is currently relatively inexpensive. However, there is nothing to indicate that it will remain so in the long term and in all countries and, in spite of improved boilers, its combustion causes relatively high pollution output. Natural gas is the least polluting fossil fuel, and it can supply heating, warm water and the energy needed for cooking. Wood can be expensive if used inefficiently, however, a high performance woodstove can prove very economical (although this may depend on its pattern of use).

In conclusion, a boiler using natural gas, or a woodstove or central heating system that is fired by wood pellets can produce effective heat that is cheaper than electric heating. The cheaper heating systems are often also the least polluting.

Seasonal efficiency of domestic boilers in the UK

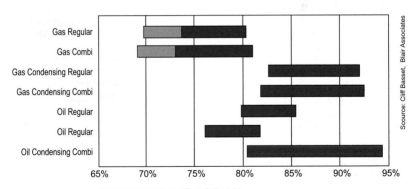

Pale orange bar shows effect of pilot light

Scource: Cliff Basset, Blair Associates

---

[2] Natural gas is not available in all areas – look in Yellow Pages for gas suppliers.

Controlling the heating

It is not enough for a system just to produce heat. Heat must be produced at the right time and at the right level. This only happens when the control and programming of the heating system is optimised. Choosing the right controls and programmers is essential to any heating system designed to save energy.

Effective heating controls reduce boiler burn and pump running times, usually by removing the 'run' command e.g. a thermostat switches off. When looking at potential controls for your heating system you should consider timers, high accuracy thermostats, thermostatic timers, optimisers and compensators. A timer can be used to reduce boiler run time, but will give poor room temperature control and will not respond to changing weather conditions. A room thermostat can be added to the system to enable it to respond to changes in weather. Remember, for every extra degree of heating the fuel consumption of your system will go up by 10 per cent. If the thermostat you fit is of low accuracy (plus or minus $2^{\circ}C$) and your thermostat is set at $21^{\circ}C$, room temperatures can dip as low as $19.5^{\circ}C$. If this is the case you are likely to reset the thermostat to $22.5^{\circ}C$ to compensate, hence increasing fuel use. Fitting your system with high accuracy thermostats/timers will save energy and cut your fuel bills. High accuracy thermostats are accurate to $0.5^{\circ}C$. Additionally, they allow different target temperatures to be set for different times. For a truly negawatt heating control system it is worth fitting advanced controllers such as optimisers and compensators. An optimiser delays the start of heating if the weather has been warm overnight. Many such systems are self-learning and will monitor the building to bring it to target

The practicalities

Does turning up a convection heater to 'high' heat a cold room more quickly?

Convection heaters, oil-filled radiators or bathroom fan heaters usually have a means for regulating temperature – for example a knob or dial marked 0, 1, 2, 3 etc up to 'max'. The room is a little cold, and you want it to heat up quickly...So you decide to push the heat up and put the dial on to 'max'. This is a bad habit to get in to: putting the heater on 'max' doesn't make it feel warmer any quicker and wastes energy.

A convection heater isn't really a form of heating in which you can vary the force of the heat, e.g. in the way that you can turn up a burner on a gas cooker. And anyway, a 1000W convection heater has only two settings: 0W (no electric current passing through), or 1000W (current running through the elements of the heater). Putting the heater on 'max' doesn't change the strength of the heater, it stays at 1000W. You are only changing the point at which the thermostat switches on or off – e.g. at 25°C rather than 19°C.

In order to boost the heat quickly, you'd have to increase the force creating the heat – to increase the wattage to 1500 or 2000W: this won't be possible without adding a second heater. Putting the thermostat on 'max' doesn't heat the room any quicker, it just makes the point at which the heat shuts off higher: if you do so, the room will not heat more quickly but will overheat. Grasp this and you'll realise that a lot of energy has been wasted for nothing.

The practicalities

Choosing temperature controls for a central heating system

Central heating can be regulated by sensing interior or exterior temperatures. Which method saves the most energy?

In the first case a simple atmospheric thermometer controls the heating of the boiler. There's no anticipation of demand for energy with regard to changes in temperature outside the house. Heating control based on exterior temperature, however, enables anticipation of likely demand and can vary the force of heating taking into account outside temperatures before they have had any noticeable effect on the temperature inside the house.

This type of control system is preferable when your radiators are at some distance from the boiler or when the house has thermal mass. It is also advisable for an underfloor heating system.

It is important to consider carefully where to put your exterior sensors/thermometers: they must be protected from direct sunlight and extreme weather conditions to function efficiently.

temperature by the set time. A weather compensator adjusts boiler flow temperature so that as the weather gets warmer, the flow temperature reduces. This improves both boiler efficiency (especially in the case of a condensing boiler) and user comfort.

Thermostatic radiator valves can be useful in areas of high incidental gains such as bathrooms and kitchens. However, they are not a replacement for a room thermostat as they do not communicate with the boiler (or the rest of your system).

Energy and pollution

Is electric heating truly non-polluting?

Solar energy is the cleanest form of energy with regard to both usage and distribution: because it is available everywhere it causes no pollution in transportation to the point of use. Natural gas is the cleanest of the fossil fuels. Its combustion generates very little SO_2, little NO_X, and, under correct combustion produces less CO_2 than coal.

Coal is by far the most polluting fuel for domestic use other than electricity: apart from substantial amounts of SO_2, NO_X, and CO_2, it releases dust, heavy metal and hydrocarbon particles. Oil and propane (or LPG – liquid petroleum gas) fall between natural gas and coal in the amount of pollution they cause in combustion and production.

Wood absorbs as much CO_2 during its growth as it emits during combustion. The global impact on the greenhouse effect of burning wood is thus small provided the regeneration of the equivalent forest is correctly regulated. But burning wood emits NO_X, and obviously has a pollution cost in transport to its point of use. When used in low-efficiency heating systems (open fires, old stoves), it can emit significant quantities of particulates and hydrocarbons. High efficiency modern systems burning wood pellets emit fewer pollutants.

There is a popular belief that electric heating is pollution free. It does not pollute the environment in use, but the same cannot be said for its production and distribution. Thus, electric heating systems (such as convection heaters, electric boilers, night storage heaters) are not completely pollution free.[3] Electricity from nuclear sources (78 per cent of the total in France in 2004, 20% in the UK in 2004 – Source: IEA) incurs a cost in nuclear waste. In the UK electricity generation in coal- and natural-gas fired power stations is a major cause of CO_2 emissions during fuel processing and plant production.

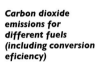

Carbon dioxide emissions for different fuels (including conversion eficiency)

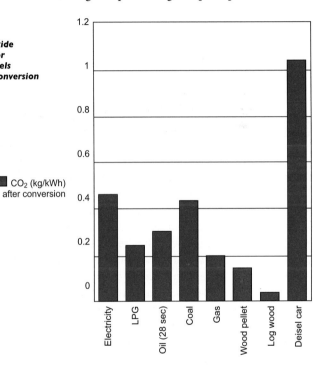

CO$_2$ (kg/kWh) after conversion

In France, more than a third of the electricity that is needed for electric heating is still produced in oil- or coal-fired power stations. In the UK, in contrast, the most common source of domestic heating is natural gas. In order to cope with increased seasonal demand in winter, a greater proportion of the required electricity is generated by these highly polluting methods increasing overall CO_2 emissions (950g per kWh produced). Even though electric domestic boilers are now around 85 per cent

---

[3] Contrary to popular belief, or rather the result of cleverly managed advertising...

efficient, the electricity that reaches them is only 30-35 per cent of the original output due to losses in generation and through delivery to the point of use.

The latest UK building regulations require all new gas and oil boilers to be condensing boilers. Boilers are also now covered by the energy labelling scheme. All condensing boilers are either band A or band B. Band B covers the range from 86 per cent to 90 per cent , and everything above 90 per cent is band A. See www.boilers.org.uk

Summary: using electric heating emits more CO_2 per kWh than oil, propane or natural gas.

Heating and global pollution

Choosing your heating system should be an ecological decision

When choosing a low-pollution heating system, it is not enough just to compare the pollution costs of the different heat sources: you also need to look at the pollution caused at each stage, from the production of the energy source to its distribution. For electricity produced by an oil-fired power station, you must take into account not only the pollution that is emitted during combustion, but also that caused during the extraction of the crude oil, the refining process and its transport to the power station. It is also important to be aware of the energy wasted during the production and distribution of electric currents, as well as the efficiency and output of any heating appliance.

This sort of exercise – comparing one fuel source against another, and its pollution cost from production to end use – allows you to compare the real impact on the environment of various methods of heating.

The most polluting form of heating is electric heating – whether via convection heaters or underfloor heating – as most electricity is generated by CO_2 emitting fossil-fuel fired power stations or by nuclear generation. Wood-fired heating[4] and solar-powered heating systems are much less polluting and by far the most environmentally friendly methods of heating your energy saving house.

The heat pump

A heat pump (also called a thermodynamic system) works according to the same principles as a refrigerator. As its name indicates, it pumps heat energy outwards via a coolant, to radiators or convection heaters.

This process obviously requires electricity to compress the coolant, but less than the volume of heat energy produced. For every 1kWh of electricity consumed by its compressor, a heat pump produces an average of 2-4kWh heat. A heat pump utilises the electricity in a way that is far more sophisticated and efficient than that of a convection heater, or an electric radiator, which produces only 1kWh of heat energy for every kWh of electricity used.

The energy source for a heat pump needs to be as constant and uniform as possible. The ground or water are good heat sources as their temperature varies little during the summer. If air is used as a heat source, then a back-up system will be needed: if the air temperature drops below 7°C the system becomes less efficient. If the air temperature drops below −5°C a back-up system will be essential.

Open fires and stoves

The sight of flames dancing in a traditional hearth is magical, but the traditional open fire is only 28-37 per cent efficient; the other 63-72 per cent of the heat generated is lost up the chimney. This inefficiency is

---

[4] Given that the felling of trees is made up for by replanting and the industry is well-managed. However, it should be noted that the burning of wood does however produce NO_x and some volatile organic compounds.

Section through an enclosed woodstove

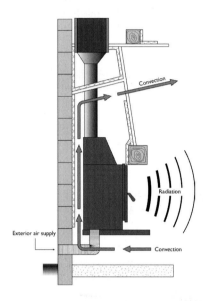

Convection

Radiation

Convection

Exterior air supply

increased when you take into account the fact that the chimney can let in draughts of cold air from outside, cooling down the internal temperature of the house.

The traditional fireplace is an atmospheric addition to the main source of heating, but modern central heating systems perform far better as space heaters.

Common in Europe, but as yet unavailable in the UK, heat recuperators can be placed directly in the hearth, slightly improving the output of an open fire (up to 25 per cent). Air recuperators suck up the air in the atmosphere, warm it up and then pump it back out. They work in the same way as natural draughts or via a mechanical ventilator.

Stoves with back boilers spread heat throughout the house with the help of a network of radiators. They are often installed alongside gas or oil fired heating systems.

Inverted combustion recuperators improve the output even more (up to 50 per cent). The aspiration of fumes and flames takes place mainly beneath the hearth.

Fitting an enclosed stove means you get to keep some of the pleasures of a wood fire while ensuring much better performance. Such stoves can be fitted in most houses as long as they fit flush into the existing fireplace. The most efficient equipment produces an output close to that of a boiler (up to 70-80 per cent). The heat is diffused throughout the house via ventilation shafts and vents. Nevertheless, these systems require regular refuelling, which often means that they become secondary heat sources. In these circumstances, wood burners can be used to help reduce energy bills when the energy supply for the principal heating system is expensive.

Information point

High performance stoves

Traditional German/Austrian ceramic stoves (kachelofen) are without doubt the best wood fired heating systems. They combine good looks with high levels of thermal comfort and up to 80 per cent efficiency. They work by integrating the burner/fire into a large masonry stove which may or may not be covered in ceramic tiles. The masonry of the stove absorbs the heat given off by the fire and then emits it over the course of several hours after the fire itself has gone out. Therefore such a stove gives off an essentially radiant, even heat, with a much reduced risk of the occupants of the house burning themselves on it. High performance ceramic stoves are usually built-in by specialist makers and are very expensive. But it is possible to source appliances which fall somewhere between a traditional cast iron wood stove and a ceramic stove or kachelofen. There are stoves available for self-assembly and fitting, from those with chimneys covered with special stone or ceramic tiles, to real ceramic/earthen stoves supplied in kit form for self-assembly.

Hints and tips

- Wood must be stored for at least two years after felling, so that its residual water content is less than 20 per cent. It must be stored under cover in a well ventilated place. Damp wood does not burn well, seriously reducing its efficiency as well as giving off increased pollutants (and potentially fatal carbon monoxide) and causing tarring of flues and chimneys.
- Hard wood (oak, beech, hornbeam, elm) is best for output of heat: 1700kWh/tonne as opposed to 1400kWh/tonne for soft wood (poplar, ash, aspen). However, the quality of the wood is just as important to both efficiency and value for money – if the wood is high in water you are paying to burn off water. Wood with a 20 per cent water content will give you more kWh/per tonne than wood with a 30 or 50 per cent water content.
- You must insulate the flue in order to prevent condensation and soot deposits. Combustible material should be at least 200mm from the inside surface of a flue or fireplace recess or 40mm from the outer surface of the chimney or fireplace recess unless it is a floorboard, skirting board, dado or picture rail, mantelshelf or architrave. Metal fixings in contact with combustible materials should be at least 50mm from the inside surface of a flue.
- There must be a vent for drawing in air in order to ensure a good draught up through the chimney and to avoid the chimney blowing back smoke. Beware of installing mechanical extract ventilation in the vicinity of an open flued appliance (such as a solid fuel stove or even gas cooker) as this can cause spillage of combustion products even if the fan is in a different room due to its lowering air pressure within the building. Ensure that the appliance is able to operate safely whether or not the fan is running (seek your installer's advice).
- The chimney must be swept at least twice a year.

High performance stoves and wood boilers

Traditional wood stoves have provided heating for generations, and many are still in use. They are generally cheap and easy to transport and to install. However, they have to be regularly refuelled, their output is often quite poor (about 40 per cent), and they emit considerable quantities of pollution, especially particulates. Numerous wood burning stoves that use out of date technology and are barely satisfactory in terms of output are still on sale. A new generation of stoves has been in use in North America and in Scandinavia for more than a quarter of a century. Thanks to technical improvements (often noted as 'clean burn'), they are up to 70 per cent efficient, and emit far less pollution. A number of models feature a large glass door allowing you to watch the flames without losing heat. Their only disadvantage is their relatively high price, but this can be recovered in a few years given that wood is an inexpensive fuel source, which can save up to 40 per cent over other energy sources.

Wood chip and pellet burners are a relatively recent innovation and there are also models available for burning sawdust and wood shavings. Wood pellet boilers are available as small as 8kW, and modulate down to about 20 per cent of maximum output. 80 per cent is the minimum efficiency to be expected, 85 per cent and above is normal so their efficiency is superior to that of the highest performance standard wood stoves, and their emissions are practically nil. Wood chip burners – only as boilers, never as stoves, and never small – are a completely different animal to pellet burners and most are not really suitable for domestic situations, being rated at 20kW+. Wood pellet boilers, rated at around 12kW, are more suitable for well insulated domestic buildings.

A store for the fuel within the wood pellet stove means it only needs to be filled every two or three days, or at worst once a day during heavy frosts. Some systems have automatic ignition. Their only disadvantages are that they need an electric lead to work a small motor, and they are expensive to buy. Wood pellets cost more than logs, and have higher calorific value because of their low moisture contents. Wood chip is only used in large systems, and is often cheaper than logs. However it has a lower calorific value because it is burnt at a much higher moisture content.

A log burning boiler feeding numerous radiators offers excellent performance with almost the same level of efficiency (85 per cent) as oil or gas burners. It is best to use an accumulator or buffer tank because log boilers run consistently at full output.

As well as pellet burners, there are also boilers that use wood chips for burning in a fuel hopper or bunker that is connected to the boiler by an auger. This is controlled with the help of a thermostat, and a chain of sparks lights the wood automatically. Management is restricted to filling the hopper every two or three days.

Solar energy

Every one of us, often without knowing it, already uses a form of solar energy!

Traditional single glazing, which works in the same way as a greenhouse, manages to recover many hundreds of kWh per year. 10-25 per cent of required heating (depending on the orientation and the characteristics of the house), can come from the sun shining through the windows of a house. However, it is worth noting that in the UK north facing single glazing will have a net energy loss over the course of a year.

Bioclimatic or passive solar design enables you to optimise the energy gained by the sun by better planning of orientation, type of glazing and thermal mass. But there are also other solar heating methods:

- a habitable veranda or conservatory attached to or integrated into the house;
- warm air ventilation systems;
- south facing external walls to most frequently used rooms;
- solar water heating systems, (often combined

with underfloor heating, though this is seldom cost effective in the UK unless also combined with a back up water heating system).

Solar heating is hence extremely flexible and can be adapted to almost any type of building provided it has suitable south-east, south or south-west facing surfaces.

A conservatory to catch the sun

A conservatory is one of the most useful elements in bioclimatic architecture. If certain rules are observed, it can reduce a house's heating requirement by 15-30 per cent and still be comfortable in summer. But watch out! A badly built conservatory can let the cold in in winter, and be too hot in summer if you don't take precautions against overheating.

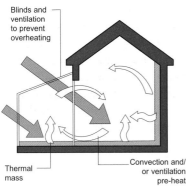

Blinds and ventilation to prevent overheating

Thermal mass

Convection and/ or ventilation pre-heat

Pat Borer/Graham Preston

Hints and tips

● A conservatory works four ways as an energy saver: it has a buffering effect against weather – the walls and windows behind a conservatory are covered from the worst of the weather, which means that the temperature difference between the house and outside is less as the conservatory's glass skin traps heat – it's a layer of insulation you can sit in; as solar ventilation preheat: fresh air entering the house will be warmed; as a lobby to external doors: reducing ventilation heat loss and acting as an airlock; it increases heat conduction to the house: the sun warms the conservatory's wall surfaces. Some of this heat warms the air in the conservatory while some migrates through interior walls to the house.

● The direction the conservatory faces is not so critical as it is for other direct solar gains such as windows. A south east aspect is ideal in the UK as the conservatory gets the early morning sun and therefore the benefits are felt throughout the day; overheating is less of a problem as it will be shaded from westerly sun at times when the outside temperature is at its highest.

● A conservatory should preferably be fitted with double low-E glazing and be unheated, with the conservatory covering as much external wall as possible, facing south east and with roof shading available.

- The walls should be of masonry construction, with an uncovered solid floor, to give it thermal mass: remember that you need to balance heat loss during 'no sun' times with the benefits of heat conduction and storage from this absorbing surface.
- The conservatory should allow for high-level ventilation for releasing heat in high summer.

Capturing the sun in the walls

Information point

Felix Trombe, the inventor of the solar furnace/kiln/ovens at Mont-Louis and Odeillo in France has also given his name to the Trombe Wall, a type of solar collector in which the heat of the sun is recovered both from the warm air circulating between the glazing and the outside face of the wall and directly from the sun's rays hitting the wall itself.

A wall designed for maximising the capture of solar energy should be solid, normally face south and have glazing fixed between 40 and 100mm from the wall's surface, making the wall a solar collector. The energy gathered in the wall is then slowly released into the house after a given delay (called de-phasing), thus allowing the house's interior to benefit from the heat that has been stored in the wall many hours after sunset. However, this depends on the thickness of the walls and the temperature: heat may be re-emitted to the outside. Any unglazed wall will still absorb solar energy but not to the maximum.

In order to be efficient at storing and de-phasing, the wall should be between 100 and 300mm thick. It can be made from concrete, brick or stone. The absorbing surface should preferably be painted black, thus capturing 90 per cent of the sun's rays. Red brick, dark green, dark blue or brown paint can also be used without losing too much of the energy[5]. 1m²

Comparison of the energy gains between different types of wall collector(A) on a house in the Ardennes

51kWh/m²/yr

With single external glazing and air circulation between the outside wall and an internal cavity wall with top and bottom vents
51kWh/m²/yr

76kWh/m²/yr

With low emissivity double glazing
76kWh/m²/yr

118kWh/m²/yr

With a transparent insulating layer of at least 5cm
118kWh/m²/yr

[5]E.g. a red brick wall (77 per cent absorption) has 14 per cent less output than a wall painted black.

of collection area on a south facing wall will heat 10m³ of air – about 30 per cent of the wall's area. In order to avoid losing heat when using shutters at night or in bad weather, it is important to install either low emissivity double-glazing or a transparent insulator of no less than 50mm. Air circulation between the wall and an insulating partition is not always necessary if single glazing is fitted and there is no transfer through any shutters at night.

The above system has been tried and tested at CAT on one of the on-site cottages and found to be unsuccessful in Welsh conditions! It is better suited to Southern European conditions. For further information see Resources.

Absorbing the sun's heat from the air

With solar powered air heating systems, the solar collectors feed a heat store, often made up of a store of rock or stone shingle beneath the house. This system's absorbing surface works out at around 1m² for 13m³ lived-in space, and about 1 tonne of shingle per m² of absorption. Air heat absorption systems have the advantage that they are usually cheaper than water systems. They are also unaffected by frost or corrosion. Air, however, is not a terribly efficient heat transfer medium as it has low thermal capacity.

There are, however, a number of disadvantages:
- The heat exchange potential of stone is eight times less than that of water and hence the exchange surface area needs to be greater
- The poor heat capacity of the air (3400 times less than water for the same volume) requires a constant airflow to maintain efficiency
- Moving air can generate dust and noise, and if the heat stores become humid, bacterial problems can develop.

For the above reasons, domestic air-based systems for houses are currently at an earlier stage of development than water based ones. On the other hand, they may prove to have a lot of potential in certain situations such as heating commercial or industrial workspaces. As yet, there have been no truly successful systems of this type installed in the UK.

Direct solar heating

Heat your floors with the sun

Today, 700 houses in France are heated by solar powered underfloor heating[6]. This is thanks to efficient technology designed for new houses, but which is also suitable for retro-fitting to existing

How a solar heated underfloor heating system works

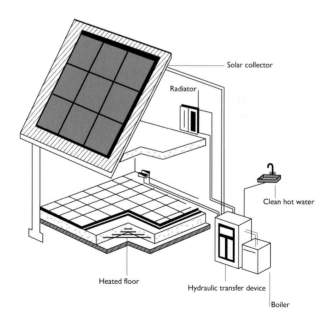

Solar collector

Radiator

Clean hot water

Heated floor

Hydraulic transfer device

Boiler

---

[6]This technique was developed by CLIPSOL who perfected the control box that runs the entire system. Remote maintenance is possible thanks to measuring equipment controllable via Minitel. More recently, GIORDANO has released a 'solar boiler' run on the same principles.

buildings during renovation. A liquid anti-freeze, heated in solar panels, is sent directly into a section of flooring through a water pipe. 130-200mm thick flooring sections both store and emit the heat produced. There is no large intermediate storage tank. Hence it is called a direct solar system. A surface area solar panel array of 120-200mm^2 for 100-150m^2 heated floor suits most situations. (These figures are for situations in Southern Europe).

A direct solar system is not recommended in the UK – here it is best to fit an alternative underfloor/water heating system that combines solar with another energy source e.g. wood or gas during the coldest months. This is because typical solar irradiation in the UK in December is around 0.5kWh/m^2/day whereas in the south of France solar irradiation is around 2kWh/m^2/day – substantially higher.

During winter days, direct solar heating can be backed up by a conventional energy source feeding into either independent heaters or the underfloor system.

Similarly, a hot water system coupled to a solar tank can preheat a hot water tank in winter, and produce hot water as required in spring and in summer.

A different kind of solar collector can be used to heat a swimming pool from April to September (in Europe), increasing its period of use by about two or three months.

Hints and tips

- For optimum efficiency solar collectors should be placed facing south at an angle of 50°
- An insulating material at least 150mm thick must be placed beneath the flooring
- Raising the heating pipes 50mm above the insulation material will improve the system's performance
- Regular checking of the pressure within the system (1 or 2 bars) and checks that controls are working satisfactorily are essential

The practicalities

What kind of additional back-up heating system should be used with solar heated floors?

The solar heated floor was originally designed with a separate, additional heating system.

Solar underfloor heating is still not used as widely as it could be and its space saving potential (not having radiators all over the house) is often overlooked.

Solar powered underfloor heating combined with a boiler (gas-, oil-, wood-fired etc) is suitable for all types of heating needs. The householder doesn't have to worry about managing two types of heating system because the regulator controls both the heat output required and which source provides what contribution to the heating required. This maximises solar gain, using the minimum power from the alternate source. It saves energy in so far as you only fire up the boiler when the heat supplied directly from the solar collector is insufficient for your space heating needs.

Solar heated floor with separate heat storage

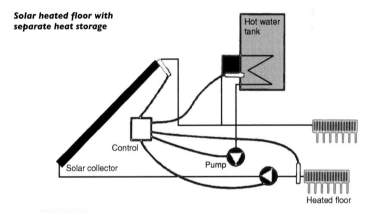

Provided that you live in an area with sufficient solar gain, combining solar powered underfloor heating with another energy source offers the greatest thermal comfort, ease of use and energy savings and it is environmentally friendly.

The disadvantage is that at present it is a relatively expensive system to fit, particularly if your space heating demand is not large (below 100m^2), but in a large house a combined heating system utilising solar water heating panels is a sensible choice for long term savings and all round comfort. In practice, in the UK solar heating is best for domestic hot water back-up for a central heating system and as a means of heating (and extending the period of use of) swimming pools.

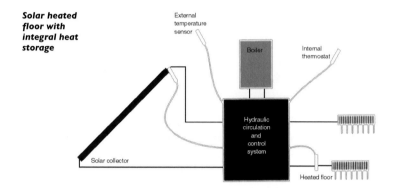

Solar heated floor with integral heat storage

External temperature sensor

Boiler

Internal thermostat

Hydraulic circulation and control system

Solar collector

Heated floor

References
(A) Study by the Ecole des Mines de Paris (Bruno Peuportier et Bernd Polster)

5. **Ventilation**
Correct ventilation

Ventilate only when it is necessary and where it is necessary

Renewing the air inside a house satisfies our requirements for oxygen and limits interior pollution by eliminating odours, fumes and toxic substances. Renewing the air also removes the water vapour produced by the occupants, and in the kitchen or bathroom avoids condensation and problems with damp.

Ventilation is vital for the safe use of boilers, stoves and other combustion devices and is crucial for maintaining a healthy indoor environment. However, in winter, when it is cold outside, heating up the cold air entering the house accounts for a substantial share of the annual energy requirement: 25-35 per cent for conventional houses, and up to 50 per cent in a very well insulated building.

In an energy saving house, therefore, correct ventilation must be as much of a priority as the prevention of heat loss through the walls. This means sorting out the following contradiction: minimising the consumption of energy without cutting down on the renewal of air.

A few simple rules can help achieve this:
- eradicate unwanted draughts,
- carefully control air flow: only ventilate when necessary,
- finally, consider energy saving solutions such as regulating air humidity, or preheating the exterior air by first filtering it through a conservatory.

The practicalities

What are the sources of air pollution in the home?

A number of products can alter air quality in your house[A]:

Domestic pollutants	Cause	Effects
Tobacco smoke	Cigarettes	Severe effect on respiratory system
Mites	Mites live in the dust and debris made up from discarded skin cells, hair, feathers, pet hair etc. They thrive in warm, humid conditions (18-22°C; >70 per cent humidity)	Their droppings (200 times their body weight over a lifetime!) are an allergen and can trigger asthma attacks in sufferers.
Pesticides and fungicides	Glue, varnish, paint, wood preservative treatments.	Respiratory problems.
Volatile organic compounds (VOCs)	Chemicals produced in the making of numerous domestic products such as paint, glue, PVC tiles, varnish, wallpaper, chipboard.	Various – depending on their concentration they can cause anything from eye irritation and respiratory problems to more serious neurological disorders.
Moulds	Condensation, humid atmosphere.	Respiratory problems. Attacks fabric of buildings: wallcoverings, swelling of wooden joinery, weakening the mechanical properties of walls.
Water vapour	Breathing: a family of four produces 8-20l of water vapour every day. Cooking, hot showers, baths. Gas heaters. Condensing tumble dryers.	Can encourage the growth of microbes. Increases the relative humidity of the atmosphere which can lead to extra condensation and discomfort. Encourages the proliferation of mites.
Oxides of nitrogen (NO$_X$)	Too much oxygen during combustion, high temperature flame.	Irritation of the respiratory system. Can cause irregularities of the circulatory system.
Carbon monoxide	Insufficient oxygen during combustion.	Takes the place of oxygen in the blood causing sleepiness, headaches, sickness, pulmonary oedema. Can be fatal.
Carbon dioxide	Breathing. Flue gas from stoves, fires, other combustion processes, leaking into living space.	In situations with high levels of CO_2 and low oxygen: nausea, migraines, breathlessness and feeling heavy-headed.
Radon	This radioactive gas comes primarily from granite and shale. It gets into the house via microscopic fissures in the flooring and tends to be a greater problem if the house has a solid floor.	Second only to tobacco as a cause of lung cancer.

A certain amount of over-ventilation may be necessary to totally eradicate pollutants (via forced ventilation in underfloor beneath suspended floor). Using 'healthy materials' containing no noxious products reduces the need for ventilation and is therefore good for those looking to save energy.

Draught prevention

Air infiltration through walls, windows and doors is often high, even in new buildings. In windy conditions air can infiltrate through the smallest of defects such as those in window joints, doors and also walls. Together, such draughts can increase energy consumption by 10-15 per cent or more.[B]

You can minimise draughts through a few simple precautionary measures. Joints between joinery and walls, walls and ceilings or walls and floors must be very carefully executed. All services should be sealed at the point at which they pass through the building envelope. A chimney, whether open or enclosed, must have an adequate air supply to work efficiently and safely. However, if the fire or stove is not in use then the chimney or flue should be closed off to prevent draughts. The kitchen extractor fan or cooker hood should process air via a carbon filter, which must be changed regularly. If the extractor fan has a flue, this will need to be equipped with a valve which only allows air to flow in one direction – out.

Finally, a tumble dryer must have its own exterior air vent.

Natural ventilation

You can ventilate naturally by creating air currents within the house. Simply open the windows or install air vents in the main rooms of the house as well as exterior vents in those parts of the house that are likely to be humid (bathroom, kitchen, WC, laundry room). Passive stack ventilation is a ventilation system usng ducts or vents from the ceiling of rooms to terminals on the roof or outdoors, which operate by a combination of the natural stack effect, i.e. the movement of air due to the difference in temperature between inside and outside, and the effect of wind passing over the roof.

Natural ventilation does not require any significant investment and does not consume any energy. Its disadvantages include insufficient control of the ventilation flows, which will depend on the wind, the prevailing weather and seasonal changes. However proper design of passive ventilation systems can overcome some of these problems.

Badly controlled, natural ventilation can cause under-ventilation and increased heating demand at either extreme. Controls such as humidistats and trickle vents are now standard in new windows, helping to prevent extremes.

Artificial ventilation

If natural ventilation is insufficient or unsuitable for your situation, controlled mechanical ventilation may be necessary. This type of ventilation is called 'mechanical' because the air is moved by an electric fan. It is 'controlled' because the operation of the fan is regulated so as to limit energy use.

More precise than natural ventilation, its disadvantages include the electricity consumption of the ventilation unit and the noise that can be generated by poorly installed units. With a simple mechanically controlled ventilator system, air vents are situated in the main rooms of the house and extractor fans in utility rooms. The renewal of air is thus constant, and neither humidity nor the number of occupants are taken into account.

In humidity controlled ventilation, exit valves modulate the ventilation flow according to occupancy and humidity. The airflow can be varied according to demand, and money can be saved on energy bills.

A two-way flow allows recovery of heat from the house's warm air: cold air coming in is warmed in an exchange with the outgoing hot air, and is then ducted into the principal rooms and finally to the outside via the utility rooms. Used predominantly in Northern Europe, this technique, however, requires buildings to have excellent air sealing, ventilation units with low fuel consumption and regular maintenance programmes.

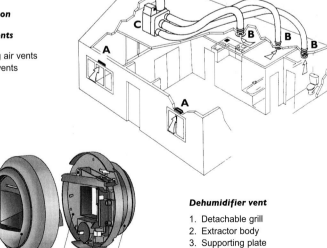

Simple ventilation system with dehumidifier vents

A Self regulating air vents
B Dehumidifier vents
C Extractor fan

Dehumidifier vent

1. Detachable grill
2. Extractor body
3. Supporting plate
4. Lip

The practicalities

Ventilation regulations

Ventilation of rooms containing openable windows on external walls
(from UK Building Regulations)

Room	Rapid/purge ventilation (opening window)	Background ventilation	Extract or passive stack ventilation (PSV)
Habitable room	1/20th floor area	8000mm^2	–
Kitchen	opening window (no min. size)	4000mm^2	30 l/second adjacent to hob or 60l/second elsewhere
Utility room	"	"	30 l/second or PSV
Bathroom (with or without wc)	opening window	"	15 l/second or PSV
Sanitary accommodation outside bathroom	1/20th floor area or mechanical extraction at 6 l/sec	"	6 l/second

*An alternative approach: the overall provision for background ventilation for the house should be equivalent to an average of 6000mm^2 per room for the rooms listed above with a minimum of 4000mm^2 in each room.

Ventilating a conservatory

Good ventilation is vital in a conservatory. The air must be circulated and renewed in order to avoid odours and humidity, especially when growing house plants. In summer, good ventilation is also essential in order to get rid of excess heat.

In winter it is also possible to use a veranda or conservatory as a heat source for pre-heating cold air coming in from outside: air enters through low vents into the conservatory, where it can recover part of the accumulated heat. The pre-heated air can then be distributed into the main rooms of the house simply by opening windows and doors leading into the house, or by opening vents in the partition wall between house and conservatory.

Such a method is simple, inexpensive and particularly effective: it can bring into the house as much energy as is lost via windows and walls.[1] However, it does lower the temperature of the conservatory, making it slightly less habitable in winter.

Air distribution into the house from a conservatory or veranda

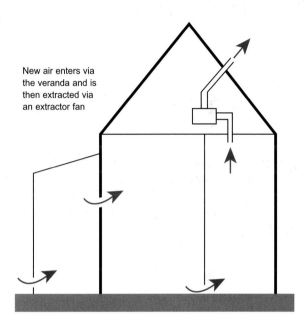

New air enters via the veranda and is then extracted via an extractor fan

References
(A) For precise figures consult *Habitat Qualite Sante* Drs S & P Deoux
(B) Olivier Sidler, *Guide des logements a faibles besoins en energie*

---

[1]This can double the 'energy production' of a veranda. Noted in *Guide des logements a faibles besoins en energie*, by Olivier Sidler, a study of design criteria for the construction of conservatories/verandas.

6. **Simply refreshing**
Cooling made simple

A house which needs air conditioning is a badly designed house

Air-conditioning cools air by blowing it through a refrigeration unit to extract the hot air and maintain an agreeable temperature and humidity.

Often presented as a symbol of progress, air-conditioning is the subject of aggressive advertising campaigns usually claiming to produce the perfect climate that should be everybody's right.[1]

In reality, having to resort to air-conditioning in Northern Europe is a sign of serious defects in the design of a house: a well planned, sufficiently insulated house, correctly oriented and with protection from the sun, shouldn't need it. Air-conditioning, which requires profligate use of electricity, has no place in the energy saving house.[2] Other simple and effective solutions exist: protecting a house sufficiently from heat, making use of cool night air, and stopping extreme variations in temperature by using the right density of building materials should be enough to keep it cool and well-aired.

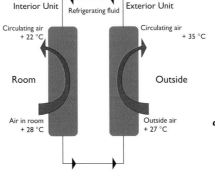

Interior Unit | Refrigerating fluid | Exterior Unit

Circulating air + 22 °C

Circulating air + 35 °C

Room

Outside

Air in room + 28 °C

Outside air + 27 °C

Principal functions of a 'split system' air conditioning unit made up of two separate elements, one situated on the inside of the house and the other outside the house

[1] EDF, Montpellier , [2] Manufacturers and installers are often reticent on this point.

Information point

Cooling floors

One technique which can be combined with underfloor heating, allows the circulation in summer of cold water through the system rather than hot. It is known as a 'reversible floor'.

In warm, humid climates, this technique is of limited usefulness: in order to avoid problems with condensation the minimum temperature at the beginning of the system must be between 19°C and 22°C[1]. This means that you will not be able to lower temperatures by more than 2-4°C.

[1]CSTB

Natural cooling

There aren't many of us who haven't been surprised by the chill felt in an old stone house in the middle of summer. In order to maintain a pleasant temperature at the height of summer, some simple rules used for generations are still valid.

First protect the house from extreme heat with architectural shields or with vegetation, by installing shutters or blinds which do not let the sun penetrate while still ensuring satisfactory light levels for daily life.

In warmer climes it's best to build dense walls, with high thermal mass, to absorb the heat of the day and to avoid interior temperatures getting too high for comfort.

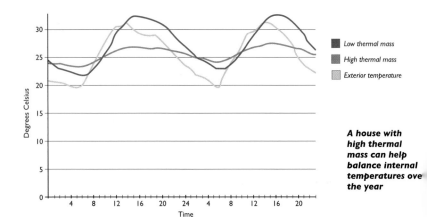

■ Low thermal mass

■ High thermal mass

■ Exterior temperature

A house with high thermal mass can help balance internal temperatures over the year

Ventilation should be mainly at night to release heat stored in the walls of the house.

Finally, it is possible to super insulate the roof[3] and exposed walls, preventing the penetration of heat at the end of eight or ten hours of sunshine.

Architectural shields

Sufficient protection from the sun can be ensured by using a whole range of fixed elements: architectural shields such as overhanging roofs, hoods, loggias, protected patios, sun shades... Nevertheless, fixed protection has its limitations. Most readily available types are only effective for south-east and south-west facing buildings. They protect only against direct radiation and not against diffuse or reflected radiation, which can make up as much as 50 per cent of solar contribution to a building's temperature. Moreover, they generate the same shade on 21 September when the weather is still warm as on 21 March, when it is often cold; therefore proving in some ways inferior to the simplest blinds or curtains, which can easily be changed according to prevailing weather conditions.

Trees, arbours, pergolas and trellis, properly sited, are good shields against solar radiation and overheating and also refresh the air through

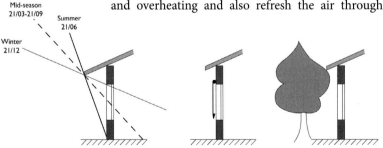

Three types of protection: fixed, portable and foliage shading

---

[3]At least the equivalent of 300mm of glasswool (R>5) for a flat roof or for those at only a slight incline.

The practicalities

Protect against overheating in summer

It is possible to use the following calculation to work out the optimum sizing of windows and roof overhang to protect against overheating in summer.

Length (L) = window height (H) in feet divided by the value F $L = \dfrac{H}{F}$

Northern latitude	F
28°	5.6–11.1
32°	4.0–6.3
36°	3.0–4.5
40°	2.5–3.4
44°	2.0–2.7
48°	1.7–2.2
52°	1.5–1.8
56°	1.3–1.5

NB Higher values give 100% shade at equinox.

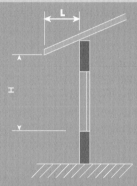

H = distance between the bottom of the window and the lowest point of the roof overhang.

L = length of the roof overhang.

East or west facing windows

For east and west facing frontages the best forms of protection against overheating from the sun are moveable systems, which will be more or less effective depending on the way they are used and where they are used.

Shading system	Shade/colour of protective system and/or application	Degree of light penetration house
Exterior slatted blinds with slats at 45° (no direct penetration by the sun's rays)	Light	15 per cent
	Dark	25 per cent
Glass fibre screen blinds	Exterior	25 per cent
Interior slatted or venetian blind	Light	55 per cent
	Medium	70 per cent
	Dark	80 per cent
Reflective curtains	Interior – silver	50 per cent
Reflective glass	Without additional protection	30-50 per cent
	With anti-glare screens on the interior	25-45 per cent

transpiration which adds oxygen and moisture to the atmosphere. This type of physical protection, unlike blinds, needs neither daily handling, nor significant maintenance and also helps to make your house a more beautiful and pleasant place to live, coping with extreme heat and maintaining good light levels in the house.

Night ventilation

Provençal wells

The 'Provençal well' pulls fresh air for ventilation into pipes buried in the ground to a depth of approx 1-2m before allowing this air into the house. In winter the ground at this depth is warmer than the outside air temperature and hence the cold air taken into the system is prewarmed during its passage through the pipe before entering the house. In summer, the ground is cooler than the outside air temperature and the well is a clever means of using the relative chill of the ground to freshen up the air coming into the house. Several such systems have been installed in the south of France to great effect.

In summer, the outside night temperature often drops below the temperature in the house. A simple and effective means of airing the house is therefore to make maximum use of night chill to dispel the heat accumulated during the day. Walls and furniture are cooled and in turn lower the overall temperature inside the house. No mechanical aid is necessary: all you need to do is make use of the cool night air by opening windows or air vents. If used jointly with building materials that have good thermal properties, night ventilation makes it possible to lower the interior temperature of the house by three to five degrees, as much as a typical air-conditioning unit might.

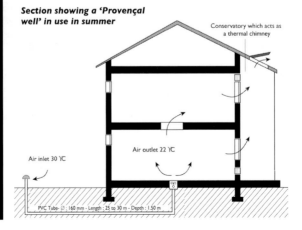

Section showing a 'Provençal well' in use in summer

Conservatory which acts as a thermal chimney

Air outlet 22 °C

Air inlet 30 °C

PVC Tube- Ø : 160 mm - Length : 25 to 30 m - Depth : 1.50 m

Daylight and artificial light

What constitutes light? Everything emits radiation depending on the level of movement of the atoms and molecules that it consists of: these produce light, and radiation. The human eye is receptive to a certain portion of the spectrum, and this enables us to see the shape and colour of the things around us.

For thousands of years, only the natural process of combustion (firewood, torches, candles, oil lamps) has been used by man to produce artificial light. During the 19th century a method of producing light from electricity was discovered and the extraordinary flexibility of electric light has had a substantial influence on our way of life ever since.

Today, lighting in the residential sector represents 10–15 per cent of UK electrical consumption. In total the UK spends £1.2 billion on electricity to light our houses. This figure could be reduced by making a few simple changes such as better use of natural light and low energy bulbs.

The European Union has proposed a ban on the manufacture of incandescent bulbs, planned to come into effect in the near future.

Lighting and natural light

Let natural light in!

All too often, the design of buildings and their interiors stops us making the best possible use of natural light and we need to switch on lights while there is still daylight outside. Carefully positioned windows, however, can provide up to one extra hour of daylight and savings of up to 50 to 100kWh of electricity per

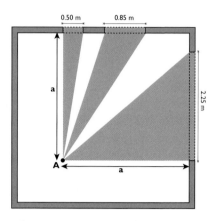

For someone sitting at point A each of the three windows lets in the same amount of daylight

year. Planning the orientation and the dimension of the windows when designing or renovating your house is therefore of great importance. Windows that let through sufficient daylight also allow us to appreciate the outside world.

There are some disadvantages of course: lighting will never be uniform throughout a building and light intensity can sometimes be too strong near windows. A distance of twice the window's height is the accepted formula for the light contribution of vertical glazing and limits light penetration to approximately 3-4m from the window.

Light from sky lights above the room is far more effective than light from a window, even on a cloudy day, because the light comes from the middle of the sky and is therefore stronger. The light available is more constant throughout the day, and gives a diffuse light with fewer shady areas. Ceiling illumination works very well in large buildings and working environments, delaying the switching on of artificial light. It also allows midsummer sun in. If placed in north facing roof surfaces sky lights let more light in than would a north facing window. On the other hand, it has a few disadvantages: there is a greater risk of water penetration and higher exposure to dust and a badly insulated opening in the ceiling can also cause significant heat loss in winter.

Gain the maximum benefit from daylight

Avoiding a few mistakes should lead to greater comfort and considerable savings:

- Avoid dark colours on ceilings – they absorb much of the light and may create an unnecessary need for artificial light
- In the kitchen, put the work surfaces beneath a window: this will avoid having to use electric light during the day for most of the year
- Install high windows, as they facilitate the entry of light
- Computer screens should be perpendicular to any nearby windows for optimal visual comfort.

> **Information point**
>
> **Natural light and the human eye**
>
> Thanks to evolution, human eyesight is perfectly adapted to working in the natural light supplied by the sun. This means that daylight will always be a more effective light for seeing in than any artificial light source.
> Natural light is non-polluting, free and visually most comfortable: make the most of it and you've got a great source of negawatts!

Artificial light

Make sure you buy lumens not watts

Replacing the sun with artificial light... The scientists who realised this dream rapidly discovered that there were two ways of exciting the atoms or the molecules of an object in order to produce visible light beams: incandescence and luminescence.

Traditional light bulbs and halogen lamps are incandescent, their light is produced by a rise in the temperature of the filament. Fluorescent tubes and low energy lamps (also called compact fluorescents – CFs – or fluocompacts) use the luminescence produced by the powder-coated lining of the tubes. Any lamp, whether incandescent or luminescent, absorbs electric power to produce light. Not all the energy, however, is converted into light and a very large proportion of available power is lost as heat. The quantity of visible light emitted by a lamp is expressed in lumens. This measures the real quantity of light provided, unlike wattage which only expresses consumption of electric power.

To compare various lightbulbs, one has to compare the amount of lumens provided against the quantity of watts consumed. This lumen/watt ratio is called luminous efficiency. The larger it is, the better the bulb is at converting energy into light.

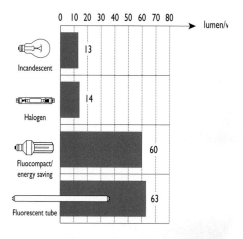

Luminous efficiency of various types of bulbs

Choosing your lightbulbs

Let's compare five types of bulb, each providing the same quantity of light…[A]

The incandescent bulb (the traditional lightbulb) is least expensive but has a very low luminous efficiency, about 13 lumens per watt. It produces little light energy (5 per cent) but lots of heat (95 per cent)!

Halogen bulbs are twice as efficient as normal incandescent bulbs, but may not be as economic as some people claim. Though very useful for indirect

	Incandescent			Fluorescent	
	Incandescent bulb	Halogen bulb	Low voltage halogen bulb	Fluorescent tube	Energy saving bulb
Luminous flow	960 lumens (lu)	840 lu	930 lu	950 lu	900 lu
Voltage	240V	240V	12V	240V	240V
Rating	75W	60W	50W	15W	15W
Luminous efficiency	13lu/W	14lu/W	19lu/W	63lu/W	60lu/W
Average lifetime	1000hrs	2000hrs	3000hrs	9000hrs	6000hrs
Replace after (given 2.7hrs use/day)	1yr	2yrs	3yrs	8yrs	6yrs
Hours of light/kWh energy consumed	13hrs	17hrs	20hrs	66hrs	66hrs
Approximate cost	£1.50-3.00	£4.00	£2.00	£3.00 (18" tube)	£0.89-13.00 (dependent on rating and quality)

Information point

Only 1.4 per cent efficient...

The electrical output of a power station (conventional or nuclear) is approximately 33 per cent. Losses across the French network are 11 per cent[1]. The output on conversion to light in an incandescent bulb is only 5 per cent... Multiply these various outputs and you get: 33% x 89% x 5% = 1.4% 98.6 per cent of the primary energy used to light up incandescent bulbs is lost as waste heat!

[1]Total losses over all uses according to the Statistiques 96/7 de la production/distribution d'energie electrique en France (DGEMP-DIGEC).

ambient lighting, halogen ceiling lamps use an awful lot of energy: a 500W halogen pencil lamp used for three hours per evening consumes 550kWh/year – more than two washing machines! However, they are generally used as spotlights. Moreover, some halogen lamps produce so much heat that they can present a fire risk.

Fluorescent tubes (sometimes called neon strip lights) are five times more economical and last eight to ten times longer than incandescent lamps. However, they are more bulky and their light is slightly less pleasant. They are most useful in places where the lighting can be of lower quality (cellar, workshop, lavatory, garage, bathroom). Lastly, low energy bulbs (CFs or fluocompacts) are four to five times more economical and last ten to twelve times longer than traditional incandescent lamps.[1] Their compact size enables them to be used directly on existing light fittings and they should be fitted wherever lights are used most.

[1]The latest generation of energy saving bulbs last for 15,000 hours (Duluxel, Osram Longlife).

Low energy bulbs

Low energy CF bulbs were produced as a result of the 1970s oil crisis, when researchers came up with the idea of folding up a fluorescent tube and integrating the choke and the ballast in the base of the lamp. Originally, these bulbs were bulky, rather heavy, and gave out a somewhat fluttering light. Since then, manufacturers have made constant technical progress, miniaturising the electronic components and improving their performance.[2]

Today's low energy bulbs provide a soft light very similar to that of a traditional incandescent light bulb, but with remarkable advantages. A 20W low energy lamp can:

- Provide as much light as a traditional 100W bulb
- Last fifteen times longer
- Release 80 per cent less heat
- Consume four to five times less energy, saving close to 1000kWh during its life span!

In the event of a change in strength of the power supply, a fluocompact bulb deteriorates only by a small amount unlike an incandescent bulb. It also retains its intensity, even if the network voltage is weak. Releasing little heat, its tube's surface reaches only 70°C instead of 230°C, which minimises lampshade yellowing and reduces the risk of fire. It can be switched on and off several hundreds of thousands of times without deteriorating.

Low energy bulbs with a separate ballast have 2 or 4 pin bases. This means you only ever have to replace the lamp, since the power supply unit invariably lasts a lot longer. They are also available with separate power supplies at low voltage, for use with solar panels, batteries or generators.

[2]By 1999 the weight of a fluocompact bulb was a quarter of what it was, efficiency improved by 24 per cent and the life expectancy of bulbs almost doubled. They now give a constant light and do not flicker. (Comparison between Philips SL 18W bulb c. 1980 and an Osram Dulux EL Longlife 15W bulb c. 1999).

A low energy lamp does not provide instant light: it takes at most 20 seconds to come on and reaches maximum illumination in one or two minutes. It may have a reduced life in situations where short term lighting is common (for example in a lavatory) and it is switched on and off more frequently.

Energy saving bulbs

12,000 hours of savings from one investment in lighting

Although a low energy bulb lasts much longer than a traditional incandescent bulb and consumes only a fifth of the energy, it is more expensive to purchase than a traditional bulb.

So is it really worth going to the trouble of changing an old bulb for a low energy one? Compare a standard incandescent 75 watt bulb and a 15 watt low energy bulb producing the same level of light.

	Ordinary bulb of 75 Watts	Low energy bulb of 15 Watts
Purchase price	£0.40	£2.00
Lifetime	1,000 hours	6,000 hours
Cost of bulbs for 6,000 hrs use	£2.40	£2.00
Cost of electricity for 6,000 hrs of use*	£52.50	£10.50
Total cost for 6,000 hrs	£54.90	£12.50
Total saving using CFL		£42.50
		* at £0.10 per kWh

In less than four months, the initial extra cost of the low energy bulb has been met, and the savings start. At the end of its life span, approx. 6,000 hours, the low energy lamp gives savings of £42.50. For an initial investment of £2.00 you will have saved £42.40 per bulb in around six years and contributed to the fight against wasted energy![3]

---

[3]There are real savings to be made by an additional initial outlay of less than £10.00 ... up to 25 per cent a year net of tax!

The practicalities

Choosing an energy saving bulb

Most energy saving bulbs last for around 6000 hours or 6 years, but some last even longer! But on energy saving grounds it is still better to buy a low energy bulb that will work for 5,000 hours than an incandescent bulb that works for 10,000 hours.[1]

A European Energy Label, similar to those for household appliances, has been designed to help people buy energy saving bulbs. They have been compulsory on all such bulbs since 1 January 2001.

The label features:

- classification of energy efficiency coded by letters from A-G, bulbs marked A being the most efficient;

- light flow in lumens(lm);

- wattage (W);

- average life expectancy of the bulb in hours, which is the average time at the end of which 50 per cent of bulbs no longer work.[2]

A European Eco-label[3] can also be found on some more environmentally friendly lightbulbs. Buying a bulb with this label attached means that you can be sure that you are buying a product which, over the course of its manufacture and disposal, has made the least impact on the environment.

---

[1]Tests (Danish study: What to Choose 1999) revealed that certain cheaper unbranded low energy bulbs have shorter life expectancies than more expensive, better quality brands.

[2]It is important not to confuse the average life expectancy of a bulb with the length of time it will actually work at its most efficient. Its useful life (CEI 64) is the average taken from 100 bulbs, at the end of which 80 per cent were still working but not producing more than 80 per cent of their initial light flow. This is less than the nominal life expectancy.

[3]To be approved under this labelling system a bulb must be efficient in terms of the light it produces, use only a little mercury and have a useful life of at least 10,000 hours. It must also be sold in a cardboard package (minimum 80 per cent recyclable) and give full details of the manufacturer.

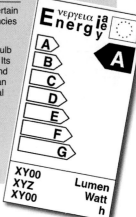

Lighting and the environment

During its lifetime, a single 20W low energy lamp saves 1000W, minimising emissions and decreasing the corresponding amount of energy production required overall. Fitting such a bulb prevents (or avoids) the emission of 60 kilos of carbon dioxide and 0.4 kilos of sulphur oxide into the atmosphere, the exact amounts dependent on which country you live in and the fuel used to produce electricity.

Reduced consumption helps save our energy resources: an oil fired power station requires 210 litres of fuel to produce 1000kWh of electricity. Therefore for every six hours of use a low energy bulb saves its own weight in oil. Low energy lamps (and fluorescent tubes) are an indisputable improvement on halogen or incandescent bulbs, however they're not completely environmentally friendly as they contain mercury, a toxic metal. The quantity of mercury used in such bulbs is nevertheless small (about 7.5 milligrams for a bulb with the European eco-label). France has classified low energy bulbs as dangerous waste since May 1999. In July 2005, the UK Government reclassified fluorescent tubes in the hazardous waste category. In the UK, there are now several companies and some local authorities offering a recycling service. The ecological assessment of low energy lamps is, all in all, much more positive[4] than for incandescent or halogen lamps which are inefficient, have a short life span and also cannot be recycled.

Information point

Recycling bulbs

In France a service for recycling fluorescent tubes is available, but only to businesses or community groups. Mercury, fluorescent powder, aluminium caps and glass – all constituents of such tubes – can now be recycled. In the UK, contact your local authority to see if they offer a service, or look online for a range of organisations offering recycling.

[4]Manufacture and waste processing only represents 10 per cent of the total ecological cost of a lightbulb. The remaining 90 per cent consists of the energy cost of transporting the bulb to the point of use, its efficiency during use and the length of its useful life.

Tips for better lighting

Here are some simple tips for better lighting at home:

- Remember to turn off the lights when you leave a room.
- Replace any standard halogen lamps or incandescent bulbs with low energy lamps. At the end of its lifetime, an incandescent bulb loses approximately 15 per cent of its luminous flow. A permanent over-voltage of 10 per cent reduces its life span by half.
- Don't watch television in complete darkness: strong light contrast increases visual tiredness, particularly between the very luminous screen of a television and the rest of the room. A low power low energy lamp, placed behind the television, should reduce this contrast with minimal energy consumption.
- A lampshade's quality and its base are almost as important as the quality of the bulb itself: a lampshade which is too dark or thick reduces the available light considerably. A good reflector diffuses a pleasant and even light where it is wanted without being blinding.
- Dust the bulbs regularly: too much dust quickly decreases the light given out.
- Switch off the electricity before changing a light bulb: simply switching off the light is not enough to eliminate all risk.
- Change a low energy lamp by holding it at the base and not by the glass tubes.

The practicalities

How to choose the right bulb for the right situation

Fluorescent lamps are often criticised on the grounds of the quality of the light they give, which some users find uncomfortable or unpleasant. These days, this criticism is no longer justified thanks to improvements in electronics and in controlling the temperature of the tube and hence the colour of the light given off.

The perceived colour of a light source is measured in Kelvins (0K = -273°C): so-called 'cold' light is generated at high temperatures and 'warm' light at lower ones.

Hence a flame burning at a high temperature is at first blue (a 'cold' colour), and then when it is cooling or about to go out, it becomes yellow-orange (a 'warm' colour).

A bulb with a colour temperature of 3500 or 4000K produces a bluer ('colder') light than a bulb with a colour temperature of 2700K:

	Power range	Luminous efficiency lumens/watt	Colour temperature	Light type type	Suggested use
Incandescent bulb	25W to 100W	11-14	1600K	Warm, intimate	DO NOT USE!
Halogen bulb 220V	60W to 500W	14-V19	3000K	Brilliant, strong	Extremely energy hungry! Do not use.
Halogen bulb 12V	10W to 50W	13-19	2900K	Brilliant, strong	Decorative or spot lighting
Fluorescent tube	10W to 60W	50-80	4000K 'white'	Brilliant, neutral	Garage, attic, cellar
Fluorescent tube			3000K 'warm white' (other colours are available)	Bright	Kitchen, bathroom
Fluorescent tube			2700K 'white gold'	Soft, intimate	Decorative lighting
Energy saving (low energy) bulbs	3W to 23W	55-65	2700K	Soft, intimate	All uses

The colour of any reflector or shade can also determine the colour of the resulting light: go for 'cold' when you want precision lighting, 'warm' when you want to create a softer light or more intimate setting.

Automatic lighting

In areas such as corridors or hallways which are often used but have a minimal lighting requirement, why not install an infrared detector? The light will be switched on when someone passes and will automatically switch off after a set time, this way the light will not stay on accidentally. A simple contact on a door can also automatically switch lights on and off in places which are often dark (a lavatory or a wall cupboard for example). If lighting is frequently needed, a linolite tube is preferable. Normal fluorescent tubes are not suited to frequent switching on and off.

Low energy bulbs are good for outside lighting or for security lights left on for long periods. In places that must stay lit all night, an interesting innovation has been developed: a lamp with an integrated twilight interruption system[5]. This energy saving lamp has cells that detect the surrounding light levels, lighting the lamp at dusk, and extinguishing it at dawn. It is possible to regulate the sensitivity of the cells in order to programme the point at which the lights will on and go off.

A fluorescent tube with an integral power supply can be equipped with an electronic ballast to control the intensity of light. This ballast functions as a dimmer.

Moreover, by coupling the tubes with an infrared detector or a photoelectric cell, and computer control, it is possible to adjust the light to precisely the desired level according to presence of people and natural light. Such devices are already in use on large-scale projects (at CAT in the Restaurant and Ateic building) and make it possible to reduce lighting costs by half: they are the lighting of future.

---

[5] DULUX EL Sensor 15W OSRAM.

The practicalities

How to vary the light from an energy saving or fluocompact bulb...

The use of a dimmer switch allows you to vary the light emitted by a bulb or lamp according to your requirements and to cut the running costs while not losing any lighting efficiency. However, unlike a number of other types of bulb, a fluocompact lamp with an integral ballast won't work with a dimmer switch.

A number of manufacturers – such as Osram, with the Dulux EL Vario, available in the UK – have developed a fluocompact lamp that can work at various light levels without the aid of a dimmer switch, which goes some way towards solving the problem.

If you switch on this bulb, it comes on at the normal maximum power rating.

But if you then switch it off and on again within 3 seconds, the power going to the luminous filament (and hence the consumption of any electric current) is reduced by half. In this way you have two choices of light level within a room without the need for any further gadgets.

References

[A]Comparison between OSRAM bulbs. Standard bulb: incandescent, clear glass Classic A FR 75. Halogen lamp 'pencil': HALOLINE® 64688. Halogen lamp TBT: HALOSTAR® Standard 64440. Fluorescent tube: LUMILUX® PLUS L18/31-830, 26mm diameter. Low energy bulb: DULUX® EL Interna.

The energy consumption of domestic appliances

How much energy do our different household appliances consume under normal conditions? A recent series of studies produced some surprising results…[1]

Beware of energy hungry appliances, such as a large ice-making refrigerator, or a halogen ceiling light. An aquarium is an energy-hungry luxury and can consume as much energy as lighting a whole house.

'h/yr

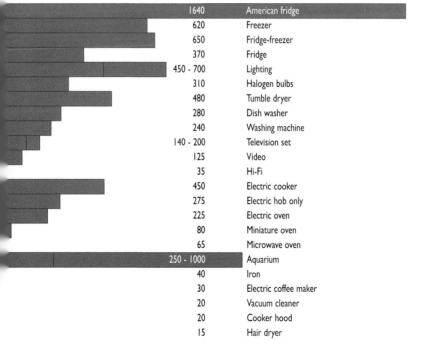

1640	American fridge
620	Freezer
650	Fridge-freezer
370	Fridge
450 - 700	Lighting
310	Halogen bulbs
480	Tumble dryer
280	Dish washer
240	Washing machine
140 - 200	Television set
125	Video
35	Hi-Fi
450	Electric cooker
275	Electric hob only
225	Electric oven
80	Miniature oven
65	Microwave oven
250 - 1000	Aquarium
40	Iron
30	Electric coffee maker
20	Vacuum cleaner
20	Cooker hood
15	Hair dryer

1 Results of the CIEL study in the Saone-et-Loire region put together by the office of Olivier Sidler for l'ADEME, EDF and the EU programme, SAVE. These results are based on the average energy consumption of the relevant type of appliance: there are, of course, wide variations between individual models.

Another thing that can waste a lot of energy without you realising it is the central heating pump, which can remain switched on 24 hours a day all year round, if the boiler is not correctly controlled by the central heating thermostat. This is often the case with old boilers or with a boiler that has not been properly installed. Badly installed pumps can consume close to 500kWh/year instead of the usual 60!

The 'Energy' label

*From A to G,
a code to follow
to the letter*

The European 'energy' label has to be displayed on refrigerators, freezers and fridge freezers, washing machines, washer dryers, tumble dryers and dishwashers. The label shows the appliance's yearly energy consumption as a letter-code indicating its energy class: A indicates low energy consumption, G extemely high consumption.[2] The 'Energy' label must also be attached to lightbulbs. Energy labelling is obligatory in the European Union and it is illegal to sell any of these appliances without correct labelling. A letter code between A and G indicates the energy performance of the apparatus. Other information also appears, such as the volume or capacity of refrigerators, the laundry capacity of a washing machine or a lightbulb's life span.

---

[2]From October 1999 companies have not been able to sell energy hungry appliances originally rated D, E, F or G in Europe. The scale A to G has been re-graded to more accurately represent the energy efficiency of the appliances currently available and now includes A+ and A++ ratings.

The practicalities

How to use energy labelling to help you choose between appliances...

Using the EU energy labelling system you can get a reasonable estimate of the total energy consumption cost of a domestic electrical appliance over its lifetime.
The calculation is a simple one[1]:
Annual consumption in kWh/yr x average cost of electricity £/kWh or Euro/kWh x estimated lifespan = total energy cost in £ or Euros
The average cost of electricity (including VAT, taxes and standing charges) is approximated in the table below:

Tariff	Use	Av cost/kWh
Charges based on use in peak hours only	Lighting	0.10
Off peak use only	Programmed appliance	0.02
	Programmed boiler	0.02
Split use between peak and off peak hours	Air conditioning	0.05
	Electric heating	0.05
	Immersion heater	0.05

Take a refrigerator selling for £150.00 and which is labelled as consuming approximately 240kWh/yr...
Estimated period of use: 15 years
Average peak electricity price per kWh: 0.10
Cost in energy during period of use: 240kWh/yr x 15 x 0.07 = £252 more than it costs to buy!
Compare the annual energy cost of two appliances of the same storage capacity made by the same company:
Category A refrigerator costs £149.95 and uses 150kWh/yr; category B refrigerator costs £129.90 but uses 208kWh/yr. The annual running cost of the category A fridge is £10.50 compared to £14.56 for the category B fridge. Hence if you buy the category A fridge you will recoup in just five years the difference in purchase price between the two machines.

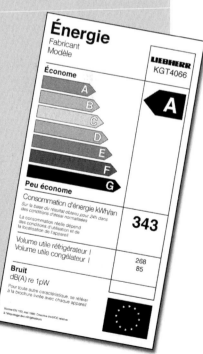

---

[1]This formula doesn't include changes in inflation or other economic factors.

Refrigerators and freezers

Information point

Refrigeration gases and the greenhouse effect...
Since the beginning of 1995 the use of CFCs (chlorofluorocarbons) has been banned. These gases have been replaced by two alternative substances: R134a (hexafluorocarbon) and R600a (isobutane). The former, however, does contribute to the greenhouse effect, and could be banned in coming years. The latter makes no discernable contribution to the greenhouse effect. It is an inflammable gas, but is used in very small quantities: a fridge-freezer contains around 50g of isobutane as opposed to an aerosol can which contains around 250g.

The refrigeration requirement of a typical household averages 400 to 1000kWh/ year. Apart from heating and hot water, refrigeration is the most significant expenditure of energy in a house. It is possible to reduce this consumption considerably by choosing a well-insulated fridge or freezer: on a standard model, 86 per cent of the efficiency loss is due to the walls letting in heat. 10 per cent is due to the heat given off by cooling food, and only 4 per cent from the infiltration of outside air and from opening the door.[3]

Instead of waiting for it to break down, replace an old machine with a very well insulated model (labelled A or B). Anything over ten years old should be replaced with an A-rated appliance. In terms of resource and energy efficiency, for a fridge bought in the USA the ideal replacement time would be after 7 years. Energy savings will result – your previous energy consumption will be reduced by a half to two-thirds. A super-insulated fridge may be 66cm wide instead of the usual 60cm, so beware: you'll have to take this into account if you're buying one for a fitted kitchen.

The moral is, beware of kitchen appliances: a large American refrigerator with an integrated ice-maker consumes three times more than a standard fridge-freezer. Such energy hungry equipment has no place in an energy saving house!

Certain appliances can make a positive contribution to energy saving (+), others will increase your consumption (-):

---

[3]Using standard 240l refrigerator with a 30l freezer compartment. Research: NEFF 397. ARENA, Zurich, 1990. Source: Manuel Ravel, OFQC Berne.a 30l freezer compartment. Research: NEFF 397.ARENA, Zurich, 1990. Source: Manuel Ravel, OFQC Berne.

	Energy advantage or disadvantage	
Super insulation (A, A+, A++ rating)	Highly insulated almost completely eradicating heat loss	+++
Interior thermometer	Permits accurate adjustment to ensure correct temperature	++
Automatic defrost	Avoids energy loss through build up of ice	+
Double compressor	Better temperature regulation. Allows either fridge or freezer to be switched off during periods of absence.	+
Frost free	Good cold distribution and low rates of ice build up, but increased level of energy consumption as a consequence[1]	– –

[1]Frost free is a particularly attractive option for those in humid areas such as the tropics, where ice forms very readily in any refrigerator.

Hints and tips

- Make sure the temperature in your fridge is neither too high (4–6°C is the maximum for storing food), nor too low (a lot of foods are spoiled if they are allowed to freeze)
- You can test the temperature of different parts of the fridge by placing a thermometer in a glass of water on each of the shelves etc
- All air vents and grills must be cleared of dust and fluff annually. Air vents must never be allowed to become blocked
- Don't position a fridge near to a heat source (oven, hotplate or hob, dishwasher or radiator), nor in direct sunlight
- Put your freezer in a cold or cool spot, but not, for example, on a balcony where it will have an increased energy demand in summer
- Look out for rapid frosting-up – a sign that your door isn't sealing properly
- If the motor seems to be on permanently it is likely that the fridge is leaking refrigerant. Get an engineer in to check and repair if necessary.
- Allow food to cool down before putting it in the fridge.
- To prevent the spread of bacteria, don't defrost frozen foods at room temperature, put them in the fridge first. This way your fridge's energy consumption rate will also benefit from the cold given off!

Should I wash up by hand or can I use a dishwasher?

For some time now, manufacturers have been reducing the water and electricity consumption of new dishwashers. So what's best for the environment and your pocket? Doing the washing up by hand or using a dishwasher?[B]

Current dishwashers use a little less water and energy than the amount used when doing the washing up by hand, but they use more detergent. Using detergents that don't contain chlorine or phosphate reduces the risk of pollution.

The most powerful machines are definitely the most economical. Some can be connected directly to a hot water source heated economically by solar collectors, or by a boiler using natural gas or wood. At a water temperature of 55°C, up to 0.9kWh can be saved by using the most energy efficient dishwasher.

	Washing up by hand	Dishwasher (av for modern machines)	High performance dishwasher with hot water fill
Number of place settings	12	12	12
Electricity consumption[1]	1.6-20kWh	1.4kWh	0.55kWh
Water consumption	30-80l	19l	15l
Washing materials	8g washing up liquid	30g detergent 3g rinsing agent 50g dishwasher salt as for standard dishwasher	

[1]AEG Oko-Favorit 808l high performance dishwasher, or the Swedish brand ASKO. These figures do not include embodied energy – the energy required to produce the machine, transport it to the point of sale or to dispose of it at the end of its life.

Ovens and cookers

Which is the best energy source to choose for the kitchen?

Transforming valuable electricity into heat for cooking raises the same concerns as those for space heating. Gas (butane, propane or natural gas) is much better adapted to the energy saving kitchen as it is much more flexible than electricity. The energy consumption of electric cookers with a ceramic hotplate is around 5 to 10 per cent lower than for those with cast iron electric plates, because a smaller mass has to be heated. Induction plates are normally 30 per cent lower, but a recent study has shown that some consume significant amounts of energy while on standby (70kWh/year).[C]

It is the way ovens and hotplates are used, however, that influences energy consumption. A microwave oven allows energy saving so long as it is used to heat up small quantities of food. It is not advisable to use it for warming up a dish that contains a lot of water or to heat food from frozen. Certain ovens have self-cleaning equipment. A catalytic system does not consume additional energy. Pyrolysis works at 500°C: it is more effective than catalysis, but consumes 3 to 5kWh each time it is used.[4]

Use a steamer or a pressure cooker to ensure you use the least energy possible when cooking.

The saucepan lid is a remarkable invention! It can save you considerable amounts of energy: to keep 1.5l of water boiling in an open pan requires around 720W of power. Put a lid on the saucepan and you'll need only 190W to keep the water boiling[D]

190 W   720 W

[4]However, a pyrolitic lining avoids the need for cleaning agents harmful to the environment.

Hints and tips

- Don't use more water than you need to boil food
  – vegetables should be just covered
- Always put the lids on saucepans
- Turn the heat right down as soon as the water starts boiling
- Never put a saucepan on a burner or hotplate that is bigger than necessary
- Switch off electric burners just before the food is cooked
  – the burner will stay hot for several minutes and continue cooking
- Never defrost food in a microwave: this is a real waste of energy and you could end up cooking it rather than defrosting it!

Washing machines

During its normal life span, a washing machine consumes an amount of electricity and water equalling the cost of its purchase. Over 10 years, the difference in consumption between two washing machines of the same capacity can vary by as much as £80.00.[5]

	Consumption: 60°C wash	Consumption over 10 years	Total cost over 10 years
Energy-hungry appliance	1.4kWh electricity/95l	3640kWh electricity/247m^3	£254.80
Energy-saving appliance	0.95kWh/50l	2470kWh/130m^3	£172.90

A high performance washing machine is the obvious choice when it comes to both water and electricity costs.

Three criteria are important when choosing your washing machine: energy consumption, water consumption, and spin cycle speed. These three factors are included on the energy label, obligatory for washing machines in Europe since January 1999.

---

[5] On the basis of 5 washes per week – 0.70/kWh electricity.

This also rates the washing and rinsing ability on a scale from A (very good) to G (poor). In addition to the energy label scheme, there is the European eco-label. This guarantees that the machine is economical in terms of energy and water (either class A or B), and that it is better for the environment because it is less polluting both in its manufacture and when in use. This eco-label is symbolised by a small flower situated on the energy label, under the code A or B.

Other alternatives exist, so what are the advantages (+) and disadvantages (-) with respect to energy and water consumption?

'Eco' option	Reduces the temperature of the wash to 40°C, but increases the soaking time	+
Variable load	Saves water by adjusting fill according to load size. It is always better to run the machine on a full load.	+
Half-load option	In principal reduces the amount of water and electricity used by 20–30 per cent. However, a cycle with a full load consumes less than two half-load cycles	=
Additional rinse option	Uses 6–7l extra water	−
Pre-wash	Increases water and electricity consumption by 15–20 per cent. Pre-washing isn't really necessary if you are using any of the more modern washing machines.	− −

A high performance washing machine is a winner on both electricity and water saving grounds...

Spin speed is a significant selection criterion, particularly if you use a tumble dryer regularly: the higher it is, the faster and therefore more economic drying your washing will be. Hence you should choose a machine with a spin speed higher than 900 revolutions/minute.

Lastly, if you have access to a source of clean hot water that has been heated by a solar water heating array, a heat pump, wood or gas boiler, choosing a machine with hot and cold water intakes can pay for itself very quickly.[6]

---

[6] Such machines are available from SIEMENS, BOSCH, AEG, FAGOR.

Hints and tips

- Always fill your machine as full as possible and don't worry about packing it in
- Clean out the filter regularly
- Where possible stick to low temperature wash cycles
- Wash at a maximum temperature of 60°C: don't go above this unless the washing is exceptionally dirty
  You will save between 1 and 2kWh per wash this way
- Use the fastest spin speed whenever possible

Information point

The real weight of washing in a fully loaded drum is very often less than 5 kilos because the washing is put in loosely and in a jumble. A bit of quick folding means you'll be able to fit more in without affecting washing efficiency.

Drying your washing

Information point

Washer-dryers are washing machines adapted to dry a half-load of washing. The drying cycle is rarely terribly efficient. Also, with certain of these appliances, not only does drying consume large amounts of energy, but also water: in condensing the water in the damp washing, the machine uses cold water. Because of this, some washer-dryers use almost as much water to dry washing as they do to wash it in the first place!

A tumble dryer uses an enormous amount of energy: drying laundry this way uses twice the amount of energy as it does to wash it! If you have the facilities, it is best to dry laundry in the open or in a sunny area: when designing or renovating a house, try to create such a space – it will save you money in the long run!

There are two kinds of tumble dryers generally available: condensation machines and extraction machines. The first consume a little more energy, but retain heat better. The second are a little less expensive to purchase, and can consume less energy, but draw in air from the room they are in, and then pump it, saturated with moisture from the washing, back out. If the room is heated then energy will almost certainly be wasted.

Laundry dries faster the greater the spin speed of the machine it is washed in: a spin speed of 500 rpm only gets rid of 35 per cent of the water content, whereas 50 per cent of the water will be removed at 1100 rpm. Since spinning washing consumes much less energy than tumble drying it, using the highest spin speed possible considerably reduces energy consumption.

Some class A tumble dryers have appeared on the market in Germany and in the Netherlands. Working in the same way as a heat pump[7], this new generation of tumble-dryers reduces energy consumption by half compared to conventional machines that function with the help of an electric resistor.

Lastly, gas tumble dryers have long been in commercial use in England and domestic machines have now been put on the market in Germany. They reduce the cost of a drying cycle by up to 60 per cent. This is a worthwhile avenue to explore if you have a domestic gas supply.

Spin speed	400rpm	800rpm	1100 rpm
Drying time Outdoors	12 hr	6 hr	5 hr 30m
Washer dryer	3 hr 30m	70m	50m
Av electricity consumption	7kWh	3kWh	2.2kWh

Information point

Generations were happy drying their washing in the sun with only a washing line and a few pegs to help. For an outlay of a few pence they got washing dry with no expense on energy, no pollution and it was beautifully fresh.
The anonymous inventor of the clothes peg should receive a prize for energy efficiency!

Hidden consumption

More and more machines remain on standby all day, either in the house (television set, video or DVD, TV decoder, hi-fi system) or in the office (computer, photocopier, printer). Others still consume a vast amount of energy, even though they appear to be switched off, like certain radios or cassettes, combined radio/cassette/CD, battery chargers for mobile telephones or various tools for DIY, etc. All this apparently insignificant electric consumption

---

[7]A thermodynamic cycle with condenser and evaporator similar to a cooling machine. Electrical energy compresses a liquid refrigerant.

passes unnoticed: such energy use is invisible[8] or appears only as a small red standby light as on a television set.

Invisible it may be, but it is still consumption: here are some examples[E]:

Television	8-20W
Video recorder	5-19W
Satellite decoder	9-16W
Satellite dish	13-15W
CD player	0-21W
TV aerial	1-2W
Radio alarm	1-3W
Radio or radio-cassette	0-3W
Battery charger	1-2W
Cordless telephone	1-6W
Telephone-fax	8-11W
Minitel 2	7W
Telephone answering machine	2-6W
Photocopier	11-25W
Computer printer (ink-jet/dot matrix)	10-20W
Computer printer (laser)	up to 50W
CRT monitor	15-30W
Laptop computer	up to 10W
Fax machine	10-20W
Halogen lamp	0-10W
Microwave oven	2-9W
Electric cooker timer	1-4W
Coffee machine	2-4W
Electric toothbrush	1-2W
Electric clock	1-3W
Video console/VDU	1W

At first sight, this type of energy consumption appears inconsequential. What are a few watts in comparison with, for example, the 1000W rating of an everyday iron?

However these machines might stay switched on 18 to 24 hours a day, 365 days a year. Their energy consumption thus makes up a considerable share of our annual electricity cost.

---

[8] For example, every time the current arrives at the initial transformer.

It is, for example, common practice to keep the following on standby all the time: television (15W), video recorder (10W), small hi-fi systems (10W), halogen lamps (6W), microwave oven (5W), and two electronic clocks (2 x 2W). The total consumption of these gadgets is 50W: in other words, they consume as much as a 50W lamp switched on night and day, and the yearly energy consumption is a whopping 400kWh![9]

400kWh is as much as a large refrigerator, nearly five times more than a very good class A refrigerator. Hence, all this apparently invisible consumption ends up having quite visible consequences: across the European Union, leaving electrical appliances on standby uses the output of eight nuclear power stations![10]

Audio-visual: beware of standby!

If you watch television for three hours per day and your television set normally works on 80W, it will consume 80W x 3h = 240Wh. If for the remainder of the day (that is 21 hours) you leave your television set on standby (15W), after having switched it off with the remote control, it will consume 15W x 21h = 315Wh...

Terrifyingly, your television is costing you more when you're not watching it if you don't switch it off at the mains!

The majority of audio-visual equipment (video recorders, decoders, satellite boxes) consumes vast

---

[9] 50W x (20–24h depending on the appliance) x 365 days.
[10] One Swiss study indicates an average of 60W standby power per house. With a hypothetical 50W permanently (440kWh/yr) on standby in the 120 million homes in the EU the total consumption amounts to at least 53 TWh. To put this into context, a 900MW nuclear power station, with a coefficient of 70 per cent produces... 900 x 365 x 24 x 0.70 = 6.7 TWh. Eight nuclear reactors (53/6.7) are required to service the standby requirements of our homes. In France, the entire output of one nuclear power station might be used solely to keep electrical gadgets on standby!

amounts of energy. A decoder permanently consumes about 17W, that is 150kWh / year, adding as much as the cost of a typical month's cable or satellite subscription to your energy bill!

Information point

It's not always easy to eliminate your use of standby mode on audio-visual equipment. For example some cheap hi-fi systems are now sold without an on/off switch. 'Switching-off' is simply a matter of putting it onto standby with a remote control. And if you unplug the system, any programming in the memory will be lost after a couple of minutes.

However, you can put switches into the cable of some equipment – mini-systems, radios, radio-cassettes – or even plug a series of machines into a multiple socket with an integral switch.

In Germany they are developing electronic devices which are able to reduce standby consumption to below 1W. Even at this low power level you are still able to switch the television on or off with a remote control. Although such devices are quite expensive to install, they are now being built into electrical and office equipment. It is worth finding out the standby load of any machine before buying and considering replacing old equipment for that with a lower standby load.

Computers, fax and photocopiers

A typical office computer (CRT monitor) consumes from 60W upwards (14" screen) and 200W (19" screen). A laptop needs much less: about 15 to 25W only. Flat screens are now remarkably efficient and consume, with basic lighting, a quarter to a third less energy than CRTs. Choosing a laptop, especially for frequent use, can save 200 to 400kWh/year.

Office computers put on 'standby' or 'sleep mode' by the user or the software system, still consume 20 to 60W: it is therefore better to switch off the computer completely, if it is going to be unused for any length of time.

With a standby consumption of about 15W, a facsimile machine wastes on average 130kWh a year simply waiting for faxes to arrive! It is important, therefore, to check a model's energy consumption before buying. When looking for fax machines and other office equipment check out US sites for models with a good energy star rating and track down the same model from a UK seller.[11] You could also try

---

[11] www.energystar.gov

the EST website. A fax-modem integrated into the office computer can lead to over consumption if the computer is left on 24 hours a day: it is then better to use an external unit with integral memory.[12]

Different types of printers have very varied energy consumption: a laser printer requires high temperatures to heat and fix the toner on paper to enable its fast print speed. The energy consumption of a laser printer in use is at least three times higher than for an ink-jet printer. If you have to use a laser printer, it is best to make sure that the machine is not left on standby all day, but is switched on and off as required.

Photocopiers work in the same way as laser printers. When they are not in service they use a lot of energy to keep the cylinder at the required heat. However, some models can be switched on quickly from an energy saving standby mode. The moral of all the above is that you should choose your model carefully and switch off whenever the appliance is not in active use.

Electrical energy consumption[F] of two types of computer printer

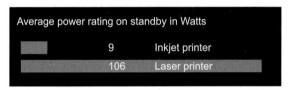

Average power rating on standby in Watts	
9	Inkjet printer
106	Laser printer

References

[A] Research undertaken by Bosch, cited in RAVEL, QFQC Berne.

[B] INFEL (Switz), ENERCO (Alain Gaumann), and Greenpeace, Belgium.

[C] Etude Ecuelle, Sidler, 1999.

[D] INFEL, ENERCO (Gaumann).

[E] From the office of Olivier Sidler, Alain Gaumann (ENERGECO, Switz), GEFOSAT measurements.

[F] Manuel Ravel, QFQC Berne.

---

[12]There are (e.g. OLITEC Smart Memory) fax-modem-answering machines with integral memory that are able to digitally record all messages that arrive when you are out. This same device is also able to find all the electronic messages that are in your in-box.

Water consumption in the home

Turn on the tap and clean water, ready to use, pours as if from an endless supply...

It is easy to forget that fresh water is one of nature's most precious resources of which there is a far from unlimited supply. 99.7 per cent of all the water on earth is salty or frozen. The rest (0.3 per cent) is either difficult to recover, or already polluted.

In the UK, we use approximately 150 litres of drinking water per person per day, of which less than 13 per cent is consumed. The rest is used for washing, laundry and cleaning. As with energy consumption, this figure is on the increase: we use twice as much water as we did in 1950.

This means that it is vital that we cut down as much as possible on any wastage. An easy task if we apply a few simple rules... To begin with, we should incorporate a few water conservation methods into our daily routines.

You can also install appliances – such as dual flush toilets or waste limiting devices – that prevent the use of more water than is really necessary. It's also important to choose with care the method by which you heat water. The point being that if you do so, you can benefit on two counts – not only reducing water bills, but energy bills too!

'We consume 8 times the water that our grandparents did'

Toilet	33%
Personal washing	25%
Drinking & food preparation	18%
Washing up	9%
Laundry	12%
Car washing & gardening	3%

Simple ways to avoid wasting water

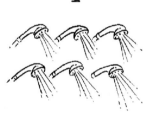

- Don't leave the tap running while brushing your teeth or shaving.
- Switch off the shower when you're shampooing or washing with soap. This will reduce your water consumption from an average of about 60 litres per shower to between 15 and 25 litres.
- On average a bath uses over 100 litres of water, compared to around 30 litres of water in a shower. However, be aware that power showers probably use more water than a bath.
- If you must water the garden, remember to do so early in the morning or late in the afternoon… and keep an eye on weather reports so that you don't water just before a rain storm!
- Wash the car with a sponge and bucket instead of a hose – a garden hose uses 1000 litres of water per hour!
- Make sure the tap is turned off after use and repair dripping taps, which can waste beween 5 and 15m^3 of water per year. A thin trickle of water from a damaged tap wastes about 10 litres of water an hour, nearly 100m^3 per year.
- Regularly check any overflow pipes on the outside of your house. If the overflow pipe is dripping it may mean a leaking ball valve in the header tank. This is easily replaced.
- Use a water butt to collect rainwater for the garden.
- Lastly, a leaky pipe can easily lose 200 to 300m^3 per year, more water than a whole family uses in a year…

The practicalities

Detecting a small leak:

- If you have a water meter fitted, check the reading at a time when no one is at home and no appliances are being used (perhaps at night when everyone is in bed, provided that no machines are programmed to work on a time delay for off peak electricity)
- Note the time and the volume of water used to the nearest litre on the dial[1] (e.g. 2305.112m³)
- Return several hours later and check the water meter again (e.g. 2305.132m³ after 5 hours)

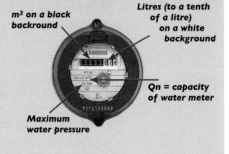

m³ on a black backround

Litres (to a tenth of a litre) on a white background

Maximum water pressure

Qn = capacity of water meter

The loss from the leak is equal to:
2305.132m³ − 2305.112m³ = 0.020m³ = 20 litres over 5 hours, that is 4 litre/hour

Common reasons for water loss:

Leaky tap	0.1 litre/h	1m³/yr
Minor drips	0.5 litre/h	5m³/yr
Dripping tap	1.5 litre/h	15m³/yr
Running tap	10 litre/h	90m³/yr
Tap left running in garden	60 litre/h	500m³/yr

[1]Certain meters will be able to give very precise figures for very small losses.

Water saving appliances

A typical tap or shower running at UK mains pressure flows at 20 litres per minute – far more water than is really necessary. A lot of clean, unused water disappears directly down the drain. The easiest way to save water without changing your basin is to fit a 'flow restrictor'. This makes it possible to maintain the water flow at 6 litres/minute instead of 20 litres/minute. This type of device can also be fitted to showers, but is not suitable for all showers, so check with the manufacturer first. Washing one's hands in running water is as effective when water runs at 6 litres/minute as when it flows at 20 litres/minute. The end result is the same but consumption is more than halved! If you want to significantly lower your water consumption, it is necessary to adapt the pressure and flow of water according to use.

If the mains pressure is higher than 3 bars, it is most effective to place a pressure reducer between the water meter and the rest of the house's water pipes. When pressure is reduced by half, flow decreases by approximately 30 per cent. However, since pressure varies in the UK and reducing pressure can result in a shower heating system cutting out, it is advisable to fit a flow reducer rather than a pressure reducer. A flow restrictor limits flow to 6 litres for taps and 12 litres for showers. (Check with your shower/tap manufacturer before fitting a flow restrictor[1].)

Another accessory makes it possible to maintain desired water speed while reducing the flow: it is attached to the ends of taps and works by aerating the water and creating a spray that is equally efficient for handwashing etc, but wastes far less water while the tap is running.[2] In the UK these are sometimes known as Tapmagic devices.

Some taps in Europe are fitted with an economy button which reduces the flow by half, but these are currently unavailable in the UK. However, it is possible to fit a lever tap: the lever is easy to push to a certain point, within this range the flow is low (eg under 5 litres/minute). For a full flow, you push the lever past a noticeable resistance point, known as the water brake.[3]

Water at the right pressure

A limiting device-regulator will adapt the flow of a shower to approximately 12 litres/minute instead of an average 18. It should preferably be placed before the hose so as to avoid over pressure. However, due to the differing water pressures common in the UK this device isn't always suitable for use since it can

[1]The Green Building Store supplies flow restrictors in the UK.
[2]If you have a wall-mounted instant gas water heater, you will need to make sure that the flow isn't allowed to drop too far: below a certain rate the flow of water out of the tap won't be sufficient to keep the heater firing.
[3]Hansa manufacture a tap with a 'water brake' and integrated adjustable flow restrictor.

cut off the water heating system – check with your system supplier or a reputable plumber before fitting any limiting device or regulator. On other types of shower, such as power showers, the propulsion of water through a pipe causes air to be taken up. This mixture of water and air reduces water consumption to 8l/minute while multiplying by four the surface of water in contact with the body.

Automatic taps that can shut off the flow after a certain period of use are also now available. Sold commercially to the trade, they are used primarily with young children or in public buildings.

But remember, households with water meters tend to use less water than those without. Water companies in the UK are obliged to fit a water meter free of charge on request! Fit one today...

Information point

Choosing water saving devices

It is best to get such things from specialist wholesalers or plumbing merchants and to buy the best you can afford, preferably those designed for applications where good water management is essential (hotels, hospitals, sports centres). Domestic fittings sold to the general public are often poorly designed, such as spray heads which are simply rubber washers.

Water-saving toilets

Toilets are the primary source of water consumption in a house. Should we really waste litres of drinking water with every flush?

Old fashioned toilets have cisterns containing up to 15 litres (though 9 to 10 litres is most common). If you have a traditional toilet, replace it as soon as possible with a modern toilet with a 6 litre cistern.[4] Some recent models, with a dual flush cistern (3l and 6l), make it possible to save water according to the flush requirement (they use up to 50 per

---

[4]For best results change the cistern, bowl and flush mechanism all together: a six litre flush won't work if it is running into an old bowl designed to work with a much bigger volume of water.

cent less water than single flush toilets). Savings are considerable: 30 to 40 m³ per year for a family of four! And if your water use is metered there will be financial savings, too. You can retrofit a variable flush lever to your cistern which interrupts the siphon during the flush cycle so the water flow stops, depending on whether you need to fully or partially clear the bowl.

Lastly, if you're unwilling to change your toilet, it is often possible to regulate the float inside the tank to limit the volume of water used or to insert a water saving device. Alternatively, fill a plastic bottle with water and place it in the cistern (see Information point).

Hot water is expensive!

Don't waste hot water and you'll save not just on water but energy too

It is possible to make significant hot water savings with no loss of comfort.

First make sure that your hot water pipes are well insulated and that they take the shortest possible route between the water-heater and the kitchen and bathroom. An uninsulated length of pipe between hot water tank and tap will contain a volume of cold water which will have to be emptied before the tap will run hot – for every 15m of pipe, you'll have to run the water for 13 seconds before it gets hot. Repeat this ten times a day and the annual cost of hot water goes up by over £20![5]

---

[5]For a 13/16mm copper pipe and a flow speed of 1.2m/s. Drinking water wasted this way is around 6.4m³/yr or £4.60 at current prices. If all the water in the pipe chills after the tap is shut off, the energy lost will have cost in the region of £18. This figure is based on losing 6.4m³ of water/year in water rates, plus electric hot water heating of 263kWh.(6.4m³ water @ £0.72/m³ plus electricity @ £0.07/kWh x 263 = £18.41).

In addition to flow regulators, several devices make savings possible:

- A mixing valve saves 10-15 per cent compared to traditional separate hot and cold taps.
- Thermostatic taps allow the water temperature to be regulated. Useful for showers, they save the water that may be lost while finding a comfortable temperature, saving up to 30 per cent on hot water.
- Finally, why not use the sun? In the UK, by collecting both direct and diffuse sunshine it is possible to heat over half the annual domestic hot water demand of a typical household.

Thermostatic tap

The immersion heater

In an electric immersion heater, the water is heated directly by an electric element submerged in an insulated tank. Even though electricity is an expensive energy, the immersion heater seems to have all the advantages: low purchase price, easy installation, few maintenance requirements, no noise, high output and the electricity used is converted into heat very efficiently. This system would be perfect but for some very serious hidden disadvantages.

First of all, electricity is an expensive form of energy. The cost of operating an immersion heater is thus considerably higher than a gas- or oil-fired water heater, even in off-peak hours.

Even if it is well insulated, a storage tank loses heat (though losses from modern tanks are very small), especially if it is located in an unheated part of the house. So, keeping the temperature of the water in the tank at the desired level requires constant energy consumption. With an immersion heater, production and use are not simultaneous: hot water is often produced unnecessarily, for example when the occupiers are away from home for a few days. Also, its capacity is limited, so you might find yourself without hot water at times when demand is high.

Lastly, as for any system that produces hot water, the electric immersion is very sensitive to limescale: if you live in a hard water area, the electric element will fur up quickly, reducing the rate of heat exchange between the element and the water.

All these losses add up, so that an electric immersion heater is really only 65 to 85 per cent efficient. But this is not all: the electricity is down to 30 per cent efficient by the time it gets to your house to heat the water compared to using a primary energy source to directly heat the water.

The total energy output of an electric immersion heater is therefore only 23 per cent, which means that on average per 4kWh of primary energy consumed, 3kWh are lost.[6] Hence heating water via electricity is hugely wasteful, a fact which shouldn't be outweighed by its convenience.

The use of electricity for the production of hot water should be kept to the absolute minimum in an energy saving house. Whenever possible, use a gas water-heater or, better still, a solar water heating system backed up by a gas-fired boiler for winter periods.

[6] This is therefore a long way off the 90 per cent achieved if you only take into account the heat exchange between the element and the stored water and not the whole production process.

Gas water heating

Gas heater

The use of gas for heating hot water is a good alternative to the electric immersion heater: gas is a relatively cheap form of energy and pollution is moderate. However the use of multipoint heating of this type is now rare in the UK.

The gas water-heater is ignited when a hot tap is turned on, so the heating is instant and ceases when the tap is turned off. Such appliances are often found in smaller properties or when there is an alternative primary energy source.[7] Gas fuelled hot water can be supplied by a combi-boiler, which also feeds the central heating system. A typical system catering for two or more people and three or more points of use in the house will have a storage tank or a tank coupled to a boiler.

Alternatively, a gas-fired hot water storage system is particularly suitable when a household needs hot water from more than one tap at the same time and when the space heating is not gas-fired. In this system, the element in the immersion tank is replaced by a gas burner, which is placed beneath the tank or immersed, which increases the output. Combustion fumes are evacuated through a chimney or a small air vent.

Information point

The pilot light – petty energy thief...

A pilot light burns somewhere around 12.5l of gas per hour.
Permanently lit it consumes approx 90m³ of gas, i.e. the equivalent of 1000kWh/yr[1].
Appliances with electronic ignition don't need pilot lights. In exchange they consume a little electricity to create a spark which lights up the boiler. The saving on gas is in the region of £20.00 a year for natural gas, and around £29.00 if you are using LPG.

---

[1] All this energy is not lost however – in winter it contributes towards heating the house provided that the heater is in the living quarters of the house.

---

[7] 8kW water heaters do not have to be fitted with a flue (in France, check Building Regulations for UK). They can be installed in any room other than a bathroom (usually the kitchen) provided that they have exterior ventilation above and below.

With a capacity of 75-300 litres and an energy range between three and 17kW, a storage tank offers the advantage of providing hot water that has sufficient flow for several taps at the same time. For the same volume of water, a gas-fired heater will raise the water temperature much faster than an electric heater: between 45 minutes and 2 hours as opposed to 6 hours.

Central heating and hot water

A central heating boiler, whether coupled to the hot water tank or not, can produce hot water. The energy source can be gas, propane, oil, wood or electricity...

Traditional boilers also produced domestic hot water, but the water remained at a permanently high temperature (at least 60°C): resulting in significant heat loss and low output.

In modern boilers, the production of hot water is independent of heating demand, boosting the output. Generally, priority is given to domestic hot water. Thanks to the power of the boiler, the tank is reheated rapidly in 10 to 30 minutes.

The tank can be integral (30 to 150 litres capacity) or it can be independent. This last system makes it possible to use one or two complementary energy sources with a single or double exchanger: you can therefore combine renewable fuel sources like wood or solar energy with conventional technology.

Hot water from the sun!

The sun can provide up to 50 to 60 per cent of the annual hot water needs of an average house in the UK, but it is not worth using the electricity produced by solar cells to heat an immersion. It is better to use the heat of the sun to heat water directly via a solar water heater.

A simple solar water heater consists of solar collectors, made of absorbent black collecting plates that are in contact with copper pipes. The output of the plates is increased when they are covered with glass (more heat is absorbed and retained, as in a greenhouse) and insulated on the back. A fluid (generally a water and antifreeze mix) circulates in these pipes. Insulated pipes connect the solar panels to a hot water storage tank. The sun heats the fluid in the panels, and this heat is then transferred to the water in the tank via a heat exchanger.

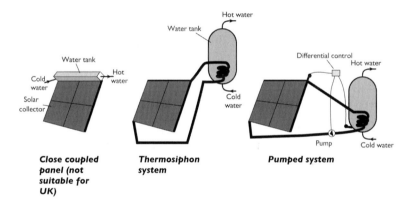

Close coupled panel (not suitable for UK)

Thermosiphon system

Pumped system

With a close coupled solar water heater the sensors and the tank are integrated into a frame used also as a support. The warm liquid in the collector is raised by the thermosiphon effect of the tank located above it. It should be noted, however, that this type of system is not recommended for use in the UK.

With a standard solar water heater the storage tank is situated in the house. Depending on the position of the collector and cylinder, two types of operation are possible:

- If the cylinder is located above the collectors, circulation can be powered by thermosiphon as in the case of close coupled systems.[8]
- If the cylinder is located below the collectors, the fluid within the collectors must be pumped around the system and controlled by a regulator.

A thermosiphoning sysetem

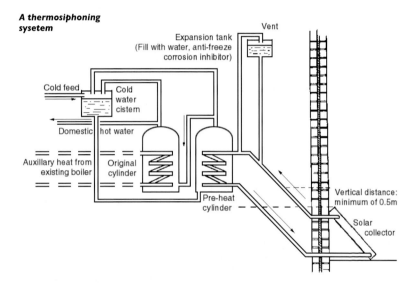

<hr>

[8] To work as a thermosiphon the solar collector/panel must be sited below the hot water cylinder. When water in the panel is heated by the sun it becomes less dense than the water in the coil half of the circuit and so it rises. As water passes through the coil it gives up its heat and becomes heavier. As long as the water coming out of the top of the panel is hotter than the water coming out of the bottom of the coil, the water will continue to circulate. There are a number of restrictions: all pipework must be at least 22mm diameter; the pipe from the collector to the tank must run uphill and the pipe back from collector to pipe downhill; the tank should be at least 1m higher than the top of the collector or more if horizontal distance is more than a few metres; serpentine pattern collectors are not suitable for thermosiphon; a thermosiphon system cannot self drain.

The practicalities

The surface area of solar collector you need for a solar water heating system

A solar array should enable you to achieve on a sunny day a water temperature of 60°C. In summer the sun is capable of meeting hot water needs for between 2 and 5 months depending on your situation. In spring, autumn and winter a solar water heating system will continue to pre-heat water even on days with minimal sunlight. A back-up energy source will be needed to bring the water up to the desired temperature.

In the majority of cases, the installation of a solar water heating system will mean putting in a solar array of between 2 and 4m². Your first calculations should be made taking into account the size of your household and your geographical situation[A]. In hot climates care should be taken not to install too large an array due to the danger of overheating in summer, diminishing efficiency and increasing expense.

If you are planning to install such a system in a house or situation which will be unoccupied for certain periods of the year, e.g. a holiday home or camp site, the orientation of your array will need to be precise and to reflect the times at which you are going to be needing hot water from it. You may need to cover the array or use a controller to prevent overheating.

Hints and tips

- In order to work efficiently, a solar water heating system must ideally have its solar collectors facing south, south-east or south-west and angled at 30° in the UK.
- It is nevertheless possible to vary these conditions without losing too much efficiency: collectors facing somewhere between south-east and south-west and inclined between 20° and 60° will work satisfactorily.
- To optimise its integration into its site a solar collector can be included in the planning of an inclined roof: better a solar array well integrated into the building at less than optimum angle than a terrible eyesore!

Choosing a solar water heater

A close coupled solar-fired water heater is less expensive than a system which has separate elements, but its aesthetic integration into the roof is difficult: it is better adapted to siting on a terrace. Its output is also weaker since the storage tank is on the exterior of the house, causing greater heat loss.

If it is possible to place the tank above the collector, then a thermosiphon system is the best solution. Its operation is not dependant on electricity and the small loss of output from the thermosiphon is counterbalanced by the savings made on the electricity consumption of the pump.

The cost depends on the systems employed and the specific constraints of the site.

In the last fifteen years many companies have started to manufacture solar collectors: early component quality was not always the best, and the fitters were not always trained.

This is no longer the case today: the technical standards laid out by the CSTB in Europe guarantee the quality and durability of solar collectors, and manufacturers ensure their fitters are properly trained. Few resources are available for testing of solar collectors in this country. Most laboratories working on testing for accreditation are elsewhere in Europe. See www.solarkeymark.org for more details. The UK standard to look for on collectors you are buying or fitting is BS EN 12975.[9] At present only two collectors available in the UK meet it. There are public databases collating information on such collectors e.g. www.solarenergy.ch, which will help with sourcing.

Information point

- The pipe network within the collector must contain an anti-freeze which is not harmful if ingested.
- It is helpful to have a thermostatic mixer in the system so that your water is at the desired temperature (50°C): water circulating in the pipes distributing it around the house will be less hot, reducing the risk of scaling, scalding and energy losses[1].
- Solar water heating systems require little maintenance. You need only to check the water pressure in the system from time to time (it should be between 1 and 2 bars), that the controls are working correctly and to check that the glass on the solar collector is clean.

---

[1] A thermostatic mixer costs around £60.00.

---

[9] BS 5918 is now out of date but some UK collectors meet this earlier standard.

The practicalities

Fitting a back-up energy source to your solar water heating system

To optimise the contribution of your solar array there are four basic rules that must be followed (B):

• capture as much of the sun's energy as you can;
• use solar energy before any other energy source;
• do not use back-up energy sources except as a complement to solar energy;
• don't mix energy sources.

These four rules seem obvious. But they are not always respected and poor connections between the back-up and the solar system can prove disastrous.

This will be the case if, for instance, the cold water is heated by the back-up energy source in advance of getting to the solar system: if there's any sun at all, you're not going to be saving money because the water will already be warm and the solar collector won't need to provide any further energy...

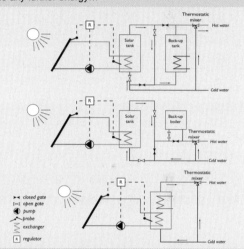

A. Back-up separated by a tank with exchanger or electric immersion heater

B. Back-up separated by an instant gas water heater

C. Integral back-up with a double exchanger tank and (or) electricity

Three system configurations are possible:

(a) The ideal system has a single cylinder with a twin coil: the first fed by solar collectors, the second fed by the back-up (see *Tapping the Sun*, available from CAT/ Resources p133).

(b) The back-up hot water system can also be heated by an instant gas water heater. In this situation the power rating of the boiler must be adjusted to take into account the temperature of the water coming in or going out and this is difficult to achieve in the UK.

(c) To keep costs and space requirements to a minimum, you can also use a combination tank (e.g. solar/electric) where the back-up heater heats only the top part of the tank, but this is technically difficult to achieve. The output, however, will be somewhat less.

Only options (a) and (b) follow all four rules.

In the UK, there are several schemes that offer training and advice for those wishing to install their own solar water heating system (see Resources).

References

[A] ADEME *paper Le chauffe-eau solaire individuel, no. 980*, 1992.

[B] *La thermique solaire, ses particularites Andre Rio,* from Cegibat no. 82.

Energy efficiency: a source of innovation

All the materials and the equipment which we have mentioned left the research laboratories a long time ago. They have been tested, and then certified. Their mass production has made it possible to gradually lower costs.[1] But in the exciting and evolving world of energy efficiency, many new innovations are now at the experimental stage.

In a few years, intelligent windows, transparent insulation, solar roofs, co-generation and more effective light bulbs will be part of our daily life.

The intelligent super-windows of the future

Intelligent windows are on their way!

The window is the thermal weak point of a house: $1m^2$ of single glazed window lets out as much heat as $10m^2$ of correctly insulated wall. Currently available insulation for walls is already very efficient, so windows are where it is particularly important to improve the level of insulation.

A German company, Pazen, distributes windows made up of double or triple low-emissivity glazing with an inert gas between the layers of glass and a thermal bridge cut in three places. The level of insulation of such windows is claimed to be exceptional. The U-values for double low-E argon hard filled glazing are equal to approximately $1.8W/m^2K$ ($1.2W/m^2K$ for soft coated) for hard coated triple low-E argon filled glazing, U-values are equal to approx $< 1.0W/m^2K$ ($0.8W/m^2K$ for soft coated) compared to approximately $0.5W/m^2K$ for 60mm

---

[1] Good quality low energy bulbs are now available for about £0.48.

polystyrene. (The double low-E argon hard filled glazing is fitted in the Ateic Building at CAT.)All these units should now be quoted to BS EN 673.

Vacuum double-glazing is a new kind of insulation glass. Vacuum double-glazing with one pane of clear glass and one pane of low-E glass, with a 7mm vacuum gap can achieve a U-value of $1.4W/m^2K$. In order to prevent the glass being crushed by the surrounding air pressure, there are spacers in the vacuum space. The seals must also be very good to maintain the vacuum. A further advantage of vacuum double-glazing is that it will not suffer from condensation or frost.

Another new concept is smart windows. Currently, this type of window has more benefits of privacy and shading, than for actual energy saving – so are more appropriate in hot countries where they may also save on air conditioning. These intelligent windows have a film with liquid crystals integrated into the glazing. An electric field directs the crystals so that the glazing can become transparent or opaque depending on light levels.

Another super-window option is to include photovoltaic cells in glazing, thereby turning the window into a producer of energy. The semi-transparent photovoltaic roof at CAT's restaurant is an example of this.

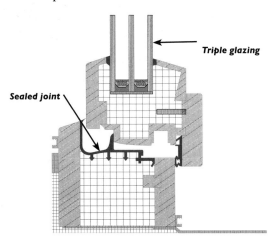

Section through a Pazen super-window

Triple glazing

Sealed joint

Photovoltaic roofs...

Solar roofs are the latest thing

A roof that transforms itself into a centre of energy production... This Jules Verne-type fantasy is already a reality for thousands of people across Europe, the US, and Japan.

The principle is simple: photovoltaic solar panels are put onto the house roof.[2] The house's electric systems are connected both to the mains and to the solar panels through an inverter-regulator, without any need for battery storage. When the sun is shining, the solar cells meet the whole or part of the house's electrical demand. On cloudy days, at night, or when the energy demand is high, the mains covers the shortfall. And if the solar cells provide more than you need, you can sell some of it back to the grid!

Solar roof at CAT

Tomorrow...cogeneration aka Combined Heat and Power (CHP)

Cogeneration simultaneously produces electricity and heat. This double system of energy production allows a total efficiency of 80-90 per cent, against the 35 per cent from thermal power stations (traditional or

[2]Solar shingles or slates serve as the actual roof covering. For suppliers contact the British Photovoltaic Association: www.pv-uk.org.uk (see Resources).

*The power
station is in the
garage*

nuclear), which discharge surplus heat. Cogeneration thus makes it possible to maximise the use of chemical energy potential. Until now, the application of this system has been confined to sites with large electricity and heat requirements. A number of different companies have researched the potential for domestic cogeneration.

The Swiss company Ecopower, with the assistance of the Engineering School in Bienne, constructed a mini-cogenerator consisting of a gas-fired thermal engine and a peripheral rotor generator. An inverter then supplies an electric current at a constant 50Hz/320V.

Heat is recovered during the cooling of the engine but also via gas exhaust. This mini-cogeneration system can provide 1 to 15kW of heat, and 0.5 to 0.7kW of electricity according to the demand for heating and electricity.

The system works in different modes, making it possible to modulate heating according to requirements. No back-up heating source is necessary, unlike previous attempts at domestic cogeneration. Any excess current produced is fed back to the mains. In Switzerland a good price is paid for this power during periods of high electricity consumption. Equipped with a catalyst and regulated by a lambda sensor, a mini-cogenerator has a very low polluting emissions rate, comparable with most modern gas-fired boilers.

Other manufacturers are about to enter the domestic cogenerator market:

- in the United States, Plug Power, a fuel cell device, has developed a 7kW generator, fuelled by gas.
- Norway Sigma Elektrotecknisk is about to market a cogenerator powered by a Stirling engine. It has the advantage of being able to be fired by any fuel, including wood and biogas.

- in Germany, Fichtel and Sachs manufacture an electric 5kW cogenerator running on gas with a diesel engine.
- BG Technology
- Victron/Wisper

Invisible insulation...

Thanks to polar bear hair!

Transparent glazing lets solar energy directly into the interior of a house. Why might an opaque wall not also collect and transmit solar radiation? This idea seemed nothing but a Utopian dream until polar bear fur (!) inspired an elegant solution...

The layout of the white hair of the polar bear's coat helps it trap valuable thermal radiation. This discovery led to the creation of transparent insulators made out of a translucent, insulating, capillary foam.

Placed in front of solid walls on the sunny side of the house, these transparent insulators let through solar radiation. This then hits a wall painted with a dark, heat absorbant colour. The heat is then transfered into the house through conduction over several hours, thus making it possible to benefit from solar heat in the evening and into the night. In turn heat radiating from the wall out towards the exterior is stopped by the transparent insulation.

The translucent aspect of the insulator masks the darkness of the absorbing surface of the wall, allowing interesting visual effects. Transparent insulators can also be used as diffusion glazing for dark rooms without the danger of dazzling.

Several houses in the Ardennes[3], in France, already use transparent insulators proving there is interest in these innovative materials.

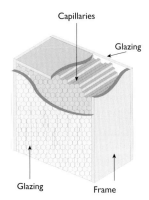

Capillaries

Glazing

Glazing   Frame

---

[3]Timber frame houses in Mouzon. Architect: Jacques Michel. Measurements: Bruno Peuportier, Centre energetique de l'Ecole des Mines de Paris.

High performance bulbs...

*Small
bulb, high
performance*

As we have seen, low energy lamps are a marked improvement on conventional incandescent light-bulbs. Further improvements in the output and quality of such lamps are still possible, for example bulbs with metal halide arc technology. This type of discharge lamp contains in addition to mercury, a component with a light efficiency of 90 lumens/watt, in contrast to the 60 of a low energy lamp.

Previously only available from 250 to 1000W, lower ratings (35-70W) are slowly coming onto the market. Compact, they have a superior output to that of the best lightbulbs or fluorescent tubes, but are no bigger than a small tungsten halogen bulb. The light given out is, however, a little colder than that of incandescent bulbs, and the bulb takes a few minutes to reach its full brightness. It is therefore unsuitable for lamps that are frequently switched on and off.

Metal halide bulb

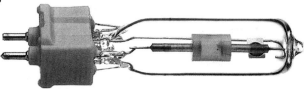

Adding up the negawatts

Throughout these pages, the aim has been to produce negawatts without sacrificing quality of life – and also to make best use of the energy available to us.

We have seen a whole range of techniques and materials for planning and building energy efficient houses. The equipment and appliances which make it possible to employ each kWh with greater effectiveness have also been examined. Taking some very simple steps can create considerable energy savings.

What are the consequences of such an approach? Let's look at the energy expenditure of one family, and then evaluate the impact on them of negawatt production.

More than 60 per cent savings!

Consider the two following situations:

– a conventional house made from traditional materials: the inhabitants do not really check their consumption,

– a negawatt house of the same size, with bioclimatic design, equipped with energy saving appliances: the inhabitants try not to waste energy but don't restrict their quality of life.

Let's compare their energy and water expenses point by point (bearing in mind that though the examples are based on the energy consumption of a house situated in a southern French climate the principle applies to any traditionally built house set against a negawatt house. The costs illustrated are those UK customers would pay if they were using these amounts of energy[1]:

Conventional House			Negawatt House		
Heating[1]	electric convection heaters	10330kWh @ £0.10= £103.30	bioclimatic design & natural gas heating	3430kWh @ £0.03 = £102.90	90% less
Air conditioning[2]	with air conditioning	2342kWh @ £0.10 = £234.20	no need for air conditioning	0kWh @ £0.10 = £0.00	100% less
Lighting[3]	conventional bulbs halogen bulbs	855kWh @ £0.010 = £85.50	Energy saving bulbs	263kWh @ £0.10 = £26.30	69% less
Domestic appliances[4]	conventional appliances	2500kWh @ £0.10 = £250.00	energy-efficient appliances	1875kWh @ £ 0.10 = £187.50	25% less
Energy consumed on 'standby'[5]	60W permanently	500kWh @ £0.10 = £50.00	30W permanently	250kWh @ £0.10 = £25.00	50% less
Cooking[6]	electric hob and oven	1184kWh @ £0.10 = £118.40	energy efficient gas cooker	964kWh @ £0.03 = £28.92	77% less
Hot water[7]	electric immersion heater	3000kWh @ £0.10 = £300.00	solar water heating panels and gas back-up boiler	3500 @ £0.03 = £105.00	67% less
Cold water[8]		240m³/yr @ £0.72 = £172.80	energy efficient appliances and careful use	144m³/yr @ £0.72 = £103.68	40% less
Totals	**£2243.90**		**£579.30**		
Total annual financial saving:			**£1664.60 74% less**		

[1]Conventional house: 100m², 250m³, coefficient G conforming to current standards, with electric convection heaters. Basic requirement: 10330kWh. Negawatt house: 100m², 250m³, south facing conservatory, bioclimatic design, supplementary heating from natural gas wall heaters. Basic requirement: 3430kWh.

[2]Conventional house: individual air conditioners that ensure maximum temperature of 27°C. Negawatt house: natural air conditioning via sun shading and night ventilation, artificial air conditioning not required.

[3]Conventional house: incandescent lightbulbs and one 500W halogen bulb. Negawatt house: 100 per cent energy saving fluocompact bulbs.

[4]Conventional house: appliances with EU Energie label rating C-D. Negawatt house: appliances with Energie label rating A or better.

[5]Conventional house: no particular precautions. Negawatt house: standby kept at a minimum.

[6]Conventional house: electric oven and hotplates. Negawatt house: gas cooker.

[7]Conventional house: electric immersion heater. Negawatt house: solar water heating panels with natural gas back-up boiler.

[8]Conventional house: 60m³ per year per person. Negawatt house: reduced to 36m³ per person per year thanks to water saving taps, high performance appliances and frugal habits!

[1]Calculations: based on a family of four living in a house of 100m². Av. electricity price £0.10/kWh; av. gas price £0.03/kWh, av. water price where metered £0.72m³

It is therefore possible to save more than 74 per cent on energy and water by better design and planning, by using economical equipment, and by consciously trying not to waste any energy.

And all this without any drop in standards, quite the opposite in fact: a negawatt house ensures better comfort levels because it has been designed and built with extreme care.

Thousands of millions of negawatts

The potential for energy saving is enormous. Saving electric energy is possible with the help of a few simple measures and by using currently available equipment[A].

For domestic and tertiary consumption alone, such a programme would save 61 billion kWh annually!

	Potential savings in TWh/yr	Equiv in no.s of 1000MW nuclear reactors
Low consumption lightbulbs	7TWh/yr	1.2
High performance office lighting	8TWh/yr	1.4
Low consumption fridges and freezers	11TWh/yr	1.9
Measures to limit the amount of electric heating	3TWh/yr	0.5
Replacement of existing electric heating	10TWh/yr	1.8
Insulation of houses and old apartment blocks	11TWh/yr	2.0
50% cut in 'standby' use	3.5TWh/yr	0.6
High performance office equipment	2.5TWh/yr	0.4
Improvements in tertiary energy usage (back-up, air conditioning etc)	5TWh/yr	0.9
Total	**6GWh/yr**	**10.7**

A wake-up call...

For years we have run our households without trying to limit the mountains of household rubbish we produce, without thinking that many of our waste products could be saved or recycled. Little by little, we are becoming aware of these wasted resources, and we are learning how to limit them.

We need to do the same with regard to energy. Fossil fuels are consumed without stopping to assess the consequences of using these cheap but finite and polluting resources – and occasional petrol supply crises are all too quickly forgotten.

Careless waste must be curbed. It is not enough just to wait and see: improving our methods of producing and using energy can only be beneficial. Everyone can help. As we have seen, on an individual level it is simple if not cost free to cut your consumption of water and energy by more than half. And governments can help by putting in hand measures to ensure that renewable energy sources are used wherever possible in preference to non-renewable ones. Otherwise, at the rate we are going, our planet will not be able to sustain the 6 billion there are of us today, never mind the 10 billion we will be tomorrow...

Each and every one of us can help by adopting the negawatts system.

References
(A) INESTENE study and Debat national sur l'energie, 1994, ADEME. Based on 5.7 TWh/yr per 1000MW reactor at 65 per cent load rate.

The Home Information Pack and Compulsory Energy Performance certificates

What it means for you

Introduction

Since 1 August 2007 all owners of residential buildings in the UK have been required to prepare a home information pack (HIP) including an energy performance certificate (EPC) when putting their home up for sale. The EU Energy Perfomance of Buildings Directive requires all properties sold or let to have an EPC, and must be complied with by January 2009.

At the time of writing the requirement applies to homes of four or more bedrooms; owners of homes of three bedrooms or less will have to adhere to this requirement once 2000 domestic energy assessors (DEA) have been registered – you will need a DEA to issue you with an accredited certificate recognised in law. (Landlords will be required to issue EPCs for prospective tenants by a date yet to be announced.) To check whether or not the required number of DEAs is in place go to www.homeinformationpacks. gov.uk, www.energysavingtrust.org.uk or call your local energy advice centre. Your local council's building control department should also have this information.

Why is the EPC a compulsory part of the HIP?

The EPC tells a home-buyer about the energy performance of the house they are looking to purchase – and will rate it between A and G (much like the current energy rating system for household

70 per cent of buyers consider energy efficiency to be an important factor when they decide on a house

appliances) for energy efficiency and environmental carbon dioxide impact. The average UK home currently achieves either a D or an E rating – largely due to the proportion of housing stock constructed more than ten years ago and with insufficient energy conservation measures either built in at the time or retrofitted thereafter.

The EU directive is designed to encourage property owners of all types to put in hand energy conservation measures. It should also encourage buyers to invest in properties that have the best possible ratings, since they will save money on future bills by doing so, they may even be prepared to pay a premium for a property with a good EPC rating. Research done by the Energy Saving Trust suggests that purchasers may be willing to pay up to £10,000 more for an environmentally friendly house, and that 70 per cent of buyers consider energy efficiency to be an important factor when they decide on a house.

Information point

The EPC gives two ratings A to G. The energy efficiency rating indicates the overall efficiency of a building: the higher the rating the more efficient it is in its use of energy. The environmental carbon dioxide impact rating gives an indication of the building's impact on the environment due to carbon emissions: the higher the rating the lower its impact. So, you should be aiming to beat the average D or E when considering home improvements to make to your property whether or not you are about to move house, but especially if you are.

What do DEAs look for and how is it measured?

All DEAs are accredited by a government-approved scheme. They are trained in the use of a system of assessment known as RdSAP (Reduced Data Standard Assessment Procedure), which was put together by DEFRA and the Building Research Establishment to enable a 'rapid yet accurate' method of working out the energy performance of an existing residential building based on a site survey. It complies with the EU Directive. It isn't appropriate for new builds, which have a different SAP to comply with.

Two different sets of data are collected dependent on whether your home is considered a 'standard' or 'non-standard' dwelling. A 'standard' house is of mainstream size, construction and style with standard heating systems (including extras such as extensions, integral garages, attic conversions and heated conservatories). A 'non-standard' dwelling is one which is considered to be 'outside the limits of what would be considered "normal"' in terms of size, construction, style or heating system, but for which an SAP can still be meaningfully calculated. Those houses for which an SAP cannot be calculated are excluded and dealt with separately (there are very few of them and it is usually because they are either very large – in excess of $450m^2$ – or they are multi-residential i.e. nursing homes or student residences). Flats and sheltered accommodation, or houses in multiple occupation (HMOs) are assessed under a different system.

The practicalities

The areas of the house assessed are related both to its structure and to its energy systems i.e. its sources of light and heat. The DEA collects data relating to its type, for instance:

- Whether it is a house, bungalow, flat or maisonette; its built form: detached, semi-detached, mid-terrace, end-terrace, enclosed mid-terrace, enclosed end-terrace; perimeter and floor dimensions; age: the earliest period being pre-1900 and the latest 2003-6, at approximately 20 year intervals.
- Whether there is a heated or unheated conservatory, how many rooms are considered habitable (i.e. living rooms, not the hall, stairs, kitchen, utility etc).
- How many floors and ceilings there are, what type of materials the building is constructed from (stone: granite or whinstone, stone: sandstone, solid brick, cavity, timber frame, system built); area of glazing and glazing type.
- Other criteria such as open fireplaces, renewable energy technologies, lighting type.
- The heating system is assessed both in terms of appliance used and fuel source, coding the boiler used by boiler ID and the SEDBUK boiler index no., and also the controls in use.
- The DEA has access to reference tables for insulation values (U-values) for wall construction, insulation materials, glazing and window area that allow comparison between most commonly occurring house types.

Hints and tips

- Make the simple lifestyle changes that conserve energy: from turning off appliances at the socket and not leaving them on standby when not in use, to fitting energy saving lightbulbs in place of incandescent bulbs (see www. energysavingtrust.co.uk or www.cat.org.uk/information for quick fixes).
- Make sure that the structural elements of your house confirm to current Building Regulations standards as applied to new build houses or renovation works. The applicable parts of the regulations are Part L and Part P. Of course, it is usually more complex and expensive to retrofit such measures, but they will pay off in terms of your EPC rating.
- When replacing your windows/glazing, insulation (not just the loft but walls and floors, too), central heating boiler etc., ensure that you conform with Building Regulations or where possible exceed the best practice recommendations. Where possible purchase items carrying the EU energy label.

Get the best possible rating

Taking the measures outlined in the chapters of this book will help achieve energy conservation levels that will increase your rating well above the national average. You can download the RdSAP v3 from the web and take a look at the elements of your home that will be examined (just enter RdSAP v3 into a search engine), or contact the Energy Saving Trust to get advice on a home energy audit that will give you a good starting point as to where you need to begin improvements along the lines of those suggested here in *The Energy Saving House.*

13. **Resources**
Organisations and institutions

General

CENTRE FOR ALTERNATIVE TECHNOLOGY
Machynlleth, Powys SY20 9AZ
Tel: 01654 705950 Fax: 01654 702782
info@cat.org.uk
www.cat.org.uk
CAT's key areas of work include renewable energy, environmental building, energy efficiency, alternative sewage/water treatment and organic growing. Provides a unique 7 acre visitor demonstration centre open 7 days a week, facilities for schools/colleges, a free information service, consultancy, public courses, publications, a mail order service, and a membership scheme.

Eco-building and design

UK BUILDING REGULATIONS
Planning Portal, G/08, Temple Quay House,
2 The Square, Temple Quay, Bristol BS1 6PN
www.planningportal.gov.uk/england/professionals/
en/1115314110382.html
The most recent versions of the Approved Documents for the fourteen technical 'Parts' of the Building Regulations' requirements are accessible from this page.

AECB (Association for Environment Conscious Building)
PO Box 32, Llandysul, Camarthanshire SA44 5EJ
Tel: 0845 4569773
graigoffice@aecb.net
www.aecb.net
Offers advice on all aspects of 'green' building. Produces a directory of environmentally friendlier building products and services available as a book,

database or on the web. Members receive a quarterly magazine 'Building for a Future'. Free list of trade members available in 'The Real Green Book'.

BRE (BUILDING RESEARCH ESTABLISHMENT)
Bucknalls Lane, Garston, Watford, Hertfordshire WD25 9XX.
Tel: 01923 664000
enquiries@bre.co.uk
www.bre.co.uk
Leading centre of expertise for construction and the built environment. Provides research, consultancy and information services. Developed BREEAM – a tool for measuring environmental impact of commercial buildings, Eco Home Points, for domestic buildings, plus other environmental impact and life-cycle assessment tools.

CONSTRUCTION INDUSTRY RESEARCH & INFORMATION ASSOCIATION
Classic House, 174-180 Old St., London EC1V 9BP
Tel: 020 7549 3300 Fax: 020 7253 0523
enquiries@ciria.org.uk
www.ciria.org.uk
Carries out research and provides information for the construction industry. Offers publications on construction, design, ground engineering and water supply and runs an environmental building forum. Publishes a recycled and reclaimed construction materials handbook.

SUSTAINABLE CONSTRUCTION
www.sustainableconstruction.co.uk
Website that provides a reference point for anyone who needs more information on how to build or refurbish in a more sustainable way. The site offers free web space to anyone with a current or completed project, product or service and aims to provide guidance and assistance to improve the environmental, social and financial performance of the built environment.

FOREST STEWARDSHIP COUNCIL
11-13 Great Oak St., Llanidloes, Powys SY18 6BU
Tel: 01686 413916 Fax: 01686 412176
info@fsc-uk.org
www.fsc-uk.org
An international, non-profit organisation set up to
provide consumers with timber and forest products
from environmentally appropriate, socially beneficial
and economically viable sources through a voluntary
certification scheme. Can provide details of FSC
certified products.

Renewable energy

ZEROCARBONBRITAIN
Centre for Alternative Technology,
Machynlleth, Powys SY20 9AZ
Tel: 01654 705958
info@zerocarbonbritain.com
www.zerocarbonbritain.com
zerocarbonbritain is a practical, low-carbon blueprint
for a sustainable future. This integrated strategy for
Britain's energy policy, provides a comprehensive
solution to the triple challenge of climate change;
peak oil (addressing energy security); and global
equity.

BRITISH HYDROPOWER ASSOCIATION
Unit 12, Riverside Park, Station Road,
Wimbourne, Dorset BH21 1QU
Tel: 01202 880333 Fax: 01202 886609
info@british-hydro.org
www.british-hydro.org
Promotes the use of water power and represents the
interests of individuals and organisations involved in
the industry.

RENEWABLE ENERGY ASSOCIATION
17 Waterloo Place, London SW1Y 4AR
Tel: 020 7747 1830 Fax: 020 7925 2715
info@r-p-org.uk

www.r-p-a.org.uk
The Renewable Energy Association was established in 2001 to represent British renewable energy producers and promote the use of sustainable energy in the UK. The REA was called the Renewable Power Association until October 2005

REA's main objective is to secure the best legislative and regulatory framework for expanding renewable energy production in the UK. We undertake policy development and provide input to government departments, agencies, regulators, NGOs and others. The British Photovoltaic Association (PV-UK) merged its membership with the REA in April 2006.

BRITISH WIND ENERGY ASSOCIATION
Renewable Energy House, 1 Aztec Row,
Berners Road, London N1 0PW.
Tel: 020 7689 1960 Fax: 020 7689 1969
info@bwea.com
www.bwea.com
Professional and trade association, concerned with all aspects of wind energy. Provides information, conferences, meetings, publications and a journal – 'Wind Directions' (quarterly). Publishes and supplies 'Best Practice Guidelines', aimed at planners. It is an excellent website resource.

ENVIROWISE
Energy efficiency hotline: 0800 585794
Envirowise delivers a valuable government-funded programme of free, confidential advice to UK businesses. This assistance enables companies to increase profitability and reduce environmental impact.

SOLAR TRADE ASSOCIATION Ltd
National Energy Centre, Davy Avenue, Knowlhill,
Milton Keynes, Buckinghamshire MK5 8NG.
Tel: 01908 442290 Fax: 01908 665577
enquiries@solartradeassociation.org.uk

www.solartradeassociation.org.uk
Aims to maintain standards within the solar water heating industry and to promote the use of solar power in general. Codes of conduct, list of members available. National enquiry centre.

Energy conservation

ENERGY SAVING TRUST
21 Dartmouth Street, London SW1H 9BP
Tel: 020 7222 0101 Freephone 0800 512 012
Fax: 020 7654 2460
www.est.org.uk
Identifies, promotes and manages energy efficiency schemes in the UK. 665577. Promotes conservation policies. Manages 52 local Energy Efficiency Advice Centres for households (freephone 0800 512 012 for your nearest EEAC). www.energy-efficiency.gov.uk. Runs lots of grant schemes to assist with installation of energy saving equipment.

NATIONAL ENERGY ACTION
St Andrews House, 90-92 Pilgrim Street,
Newcastle-Upon-Tyne, Northumberland NE1 6SG
Tel: 0191 261 5677 Fax: 0191 261 6496
info@nea.org.uk
www.nea.org.uk
Provides advice on heating and insulation problems to low-income households. Offers basic training in energy conservation. Publishes Energy Action Journal quarterly. Also does research and pilot practical projects in energy efficiency.

NATIONAL ENERGY FOUNDATION
The National Energy Centre, Davy Avenue,
Knowlhill, Milton Keynes, Bucks. MK5 8NG
Tel: 01908 665555 Fax: 01908 665577
info@nef.org.uk
www.nef.org.uk
Charity promoting the efficient, innovative and safe use of energy. Promotes energy efficiency through

the NHER – National Home Energy Rating Scheme. Provides a free Survey by post to homeowners. Runs the NEF Renewables scheme including the Wood Fuel Information Service. Runs a self build solar water heating project salled Sefsol. Secretariat for the Solar Trade Association.

Water

ENVIRONMENT AGENCY
National Customer Contact Centre,
PO Box 544, Rotherham S60 1BY
Tel: 08708 506506
enquiries@environment-agency.gov.uk
www.environment-agency.gov.uk
Executive arm of the government which aims to provide a comprehensive approach to the protection and management of the environment by combining the regulation of land, air and water. Excellent factsheets and information on their website.

THE INTERNATIONAL WATER
ASSOCIATION (IWA)
Alliances House, 12 Caxton St., London SW1H 0QS
Tel: 020 7654 5500 Fax: 020 7654 5555
water@iwahq.org.uk
www.iwahq.org.uk
A global network of water professionals, spanning the continuum between research and practice and covering all facets of the water cycle.

Trade associations and accreditation schemes

ASSOCIATION OF ELECTRICITY PRODUCERS
1st Floor, 17 Waterloo Place, London SW1Y 4AR
Tel: 020 7930 9390 Fax: 020 7930 9391
enquiries@aepuk.com
www.aepuk.com
Trade associates for electricity producers and suppliers in the UK, including renewable energy.

BRITISH WATER
1 Queen Anne's Gate, London SW1H 9BT
Tel: 020 7957 4554 Fax: 020 7957 4565
info@britishwater.co.uk
www.britishwater.co.uk
Lead trade association for water and waste water industry in UK and overseas. Includes specialist organisations, civil and process contractors, management, engineers, consultants, researchers, manufacturers, suppliers, etc. Collaborates with British and European agencies in development of standards affecting the water industry. Manages a technology transfer club 'The Environmental Technology Club', for the development and application of sustainable environmental treatment technologies.

CHEAPEST GAS AND ELECTRICITY UK
www.cheapest-gas-electricity.co.uk
A website that compares the prices of gas and electricity suppliers around the UK.

FEDERATION OF ENVIRONMENTAL
TRADE ASSOCIATIONS
Henley Rd., Medmenham, Marlow, Bucks SL7 2ER
Tel: 01491 578674 Fax: 01491 575024
info@feta.co.uk
www.feta.co.uk
The recognised UK body which represents the interests of manufacturers, suppliers, installers and contractors within the heating, ventilating, refrigeration & air conditioning industry. It is split into five principle Associations:
- Building Controls Industry Association (BCIA)
- British Flue & Chimney Manufacturers Association (BFCMA)
- British Refrigeration Association (BRA)
- Heating, Ventilating and Air Conditioning Manufacturers Association (HEVAC)
- Heat Pump Association (HPA)

FOREST STEWARDSHIP COUNCIL
(see above)
Accreditation for sustainably produced timber.

CONNAUGHT GASFORCE
Connaught House, Pynes Hill,
Rydon Lane, Exeter EX2 5T2
Tel: 0870 145 3355
enquiry@connaughtgasforce.com
www.connaughtgasforce.com
Provides a range of gas services to commercial,
industrial, and public service premises.

GLASS AND GLAZING FEDERATION
44-48 Borough High Street, London SE1 1XB
Tel: 0870 042 4255 Fax: 020 70870 042 4266
info@ggf.org.uk
www.ggf.org.uk
Disseminates information and advice and provides
contracting services, membership, newsletter,
promotional campaigning. Will answer written
enquiries.

LP GAS ASSOCIATION
Pavilion 16, Headlands Business Park,
Salisbury Road, Ringwood, Hampshire BH24 3PB
Tel: 02476 711601
mail@lpga.co.uk
www.lpga.co.uk
Trade association for UK suppliers of liquid petroleum
gas (LPG) for domestic use (including cooking,
heating and hot water). Website details potential
uses and constraints of LPG, member suppliers and
installers.

WATER UK
1 Queen Anne's Gate, London SW1H 9BT
Tel: 020 7344 1844 Fax: 020 7344 1853
info@water.org.uk
www.water.org.uk

Trade association of the United Kingdom's water industries. Provides a forum for technical expertise and information sharing.

Manufacturers and suppliers

AES Ltd.
AES Building, Lea Rd., Forres, Morayshire IV36 1AU
Tel: 01309 676911 Fax: 01309 671086
info@aessolar.co.uk
www.aessolar.co.uk
Manufacturer of flat plate solar water heating panels. Roof intergration or retrofit. Pressurised and drainback systems. Fully installed or DIY kits. (Buildings and swimming pools). System design on autocad applications. Range from small and multiple, domestic to large commercial. STA member.

TAPMAGIC and other water conserving devises
Available from the Centre for Alternative Technology (see above)

CORNWELL HEAT Ltd.
Bells Lane, Hawstead,
Bury St Edmunds, Suffolk IP29 5NW
Tel: 01284 386447
cornwellheatltd@btopenworld.com
Designs and manufactures the Ecovatic wood pellet boiler (up to 50kW). And is distributor for the Sideros domestic warm air wood pellet stove (up to 25kW) and the EcoTre pellet manufacturing equipment (which processes various types of material including municipal waste and wood into a pellet fuel). Member of British Biogen.

ECO-HOMETECH (UK) LIMITED
Unit 11e, Carcroft Enterprise Park, Carcroft,
Doncaster, S Yorks DN6 8DD
Tel: 01302 722266 Fax: 01302 728634
sales@eco-hometec.co.uk
www.eco-hometec.co.uk

UK manufacturer of solar compatible combination boilers including twin coil & Ecosol Solar Heat flat plate collector with chromium/ nickel stainless steel surface for domestic systems and swimming pools. Design & consultancy service offered. Supplier of solar heating systems compatible with combination boilers.

FILSOL Ltd.
15 Ponthenri Industrial Estate
Ponthenri, Llanelli, Camarthenshire SA15 5RA
Tel: 01269 860229 Fax: 01269 860979
info@filsol.co.uk
www.filsol.co.uk
Manufactures a range of solar products, including flat plate thermal solar panels and monocrystalline silicone cell PV modules. Also designs alternative energy and heat recovery systems. Supplies, but doesn't manufacture, panels for swimming pools. Solar energy project consultancy. Solar Trade Association member. Best Buy for Ethical Consumer buyer's guide for April/May 2001. Also conducts training in designing and installing solar water heating.

GREEN BUILDING STORE
Heath House Mill, Heath House Lane,
Bolster Moor West Yorkshire HD7 4JW
Tel: 01484 461705 Fax: 01484 653765
info@greenbuildingstore.co.uk
www.greenbuildingstore.co.uk
Supplies products which promote healthier and more environment-friendly buildings. Whether specifying for a large development, or renovating your own home, they can provide cost-effective solutions.

THE HOT SPOT
53-55 High Street, Uttoxeter, Staffordshire ST14 7JQ
Tel: 08452 606404 Fax: 01889 567 625
sales@thehotspot.co.uk

www.thehotspot.co.uk
Supplies 'Relax' wood burning stoves for wood, shavings, sawdust etc with an output of 4kW to 20kW. Also supplies 'Popular' stoves for solid fuels and woodchips with output of 4kW to 6kW.

LOG PILE
www.logpile.co.uk
Webpage run by NEF renewables to enable you to source English wood pellets, chips, charcoal and logs by entering your postcode. Also has equipment suppliers and a lot of information about various wood fuels.

PHILIPS LIGHTING
The Philips Centre, Guilford Business Park,
Croydon, Surrey GU2 8XH
Tel: 01293 776675
www.philips.com
Large company making lighting products, including energy efficient light bulbs. On-line ordering facilities.

POWERTECH SOLAR Ltd.
21 Haviland Road, Fernown Industrial Estate,
Wimbourne BH21 7RZ
Tel: 01202 890234
sales@solar.org.uk
www.solar.org.uk
Designs renewable energy systems and carries out consultancy for energy and water supplies. Manufactures 100W to 6kW wind turbines and both evacuated tube and flat-plate solar water heating systems as well as hybrid systems for combined heat and power. Designs various hot water tanks for both domestic hot water and under-floor heating. Also stocks many components used for heat recovery, rain water harvesting, waste water recovery, low energy lighting, recycling and various solar powered products.

RENEWABLE HEAT AND POWER Ltd.
Pinkworthy Barn, Oakford,
Tiverton, Devon EX16 9EU
Tel: 01398 351166 Fax: 01398 351115
admin@rhpl.co.uk
www.rhpl.co.uk
Supplier of biomass fuel boilers (Veto wood chip boilers and Ecotech AB Pellet boiler). Its commercial arm, Western Wood Pellets Ltd., supplies wood pellets. Member of British Biogen.

SUNDWEL SOLAR Ltd.
1 Tower Road, Glover Industrial Estate,
Washington, Tyne and Wear NE37 2SH
Tel: 0191 416 3001 Fax: 0191 415 4297
solar@sundwel.com
www.sundwel.com
Specialises in solar water heating (flat plate systems), offers a design service, supply of materials and installation instructions. DIY kits. Manufactures and supplies direct to the construction industry. Sole UK trader for Resol controller.

WELSH BIOFUELS Ltd.
32 Chilcott Avenue, Brynmenyn Industrial Estate,
Brynmenyn, Bridgend CF32 9RQ
Tel: 01656 729714 Fax: 01656 728178
martin@welsh-biofuels.co.uk
www.welsh-biofuels.co.uk
Manufactures wood pellets boilers and an 11kW Enviro Fire pellet stove. Developing and expanding the range of boilers available. Supplies Passat pellet or chip burning boilers. Designs and installs pellet wood heating systems. Manufactures and supplies pellets.

CHP and heat pumps

COMBINED HEAT & POWER ASSOCIATION
Grosvenor Gardens House, 35-37 Grosvenor
Gardens, London SW1W 0BS

Tel: 020 7828 4077 Fax: 020 7828 0310
info@chpa.co.uk
www.chpa.co.uk
Aims to encourage the widespread use of both
district heating and combined heat and power (CHP)
in the UK. Member of British Biogen.

THE CHP CLUB
c/ AEA Technology Plc. Future Energy Solutions,
Gemini Building PO Box 200, Harwell, Didcot,
Oxon OX11 0QJ
The Carbon Trust Energy Helpline: 0800 085 2005
www.chpclub.com
Offers information on large scale industrial combined
heat and power (CHP), and small scale packaged
CHP. Also produces a quarterly publication. Full
access is restricted to members but many areas of
website available to CHP Club Browsers (register
online). Website and freephone number offering
up to 3 days free consultancy from the independent
consultants panel.

HEAT PUMP ASSOCIATION
2 Waltham Court, Milley Lane,
Hare Hatch, Reading, Berks RG10 9TH
Tel: 01189 403416 Fax: 01189 406258
info@feta.co.uk
www.feta.co.uk
The HPA is composed of leading manufacturers
and the energy utilities in the field of heat pump
technology. Any of its members will be pleased to
help with queries or problems.

JOHN CANTOR HEAT PUMPS
Llanwrin, Machynlleth, Powys SY20 8QH
john.cantor@heatpumps.co.uk
Designs, manufactures and installs water source
heat pumps, including hydro and electrically driven
systems.

Future developments

INTEGER INTELLIGENT & GREEN Ltd.
Building 9, Bucknalls Lane,
Garston, Watford WD25 9XX
Tel: 01923 665950
alankell@iandi.ltd.uk
www.integerproject.co.uk
The Integer house was built at the Building Research Establishment using construction innovation, intelligent technologies and environmental techniques. It was designed to be a home of the future, featuring lots of high tech gadgets as well as solar water heating, heat pumps, etc.

ndangered species...

ct now and help us freeze climate change.

no joke. We're already seeing the
cts of climate change. The time for
ng is over. It's time to act.

e we can see differences in our
sons here in the UK, climate change
tting developing countries hardest.
ds and droughts are becoming more
more common. People are losing
livelihoods and lives as the weather
es from one extreme to the other.

we have the chance to do something.
the help of people like you, who
, we've already got the Government
mmit to tackling climate change
troducing a law to cut UK carbon
de emissions.

with your support, we can make
hit the annual cut in emissions

we need from now to stop catastrophic
climate change.

This is our best chance to slam the brakes
on rising global temperatures. We have
the solutions. We just need your support
to make the Government implement them
immediately. Not next year. Now.

Responding to this ad is the surest way
you can help us get the result we all need.
All we ask for is £3 a month, or a one-off
donation of whatever you can afford.

Donating is easy.
Simply call FREEPHONE 0800 581 051
or visit foe.co.uk/supportus

Or you can text EARTH to 80231 for more
information and a free mobile wallpaper.